Contents

SEXISM & SCIENCE

EVELYN REED

PATHFINDER PRESS
New York

The following articles were printed first in the *International Socialist Review* and have been reprinted here with permission: "Sociobiology and Pseudoscience," October 1975; "Lionel Tiger's 'Men in Groups': Self-Portrait of a Woman-Hater," March 1977; "Anthropology and Feminism: An Exchange of Views," March 1975; "Evolutionism and Antievolutionism" (originally titled "Anthropology Today"), Spring 1957; "An Answer to 'The Naked Ape' and Other Books on Aggression" (originally titled "Is Man An Aggressive Ape?"), November 1970.

Library of Congress Catalog Card No. 77-92144
ISBN: paper 0-87348-541-6 / cloth 0-87348-540-8
Manufactured in the United States of America

First edition 1978
Second printing 1981

Pathfinder Press
410 West Street, New York, NY 10014

Introduction

Science, by definition and tradition, is supposed to be totally objective and free of bias. However, this is the ideal, not the reality.

The activities and theorizing of the scientific community do not proceed in a vacuum. They are subject to all the biases current in the established social system; these affect, and sometimes warp, their conclusions.

The influence of prejudice tends to be strongest in those branches of science that are closest to human life and its history and values. Among them are biology, sociology, anthropology, and the two younger sciences called sociobiology and primatology. These are the disciplines discussed in this book.

The essays in this compilation do not disparage the genuine discoveries and advances made by workers in these various fields. They seek rather to show in what specific respects the infiltration of pseudoscientific notions distort the facts and obscure the truths to be found in them.

Much has been written in recent years about the racism to be found in the conclusions of certain geneticists and other writers. Less attention has been paid to the presence of sexist stereotypes in the biological and social sciences dealt with in this book. Some of these are being brought to light by partisans of women's liberation, who are more likely to be sensitive to them and aware of the harm they do both to the pursuit of scientific truth and the cause of social progress.

The first three essays in this book are primarily concerned with the newer sciences of sociobiology and primatology, the last five with the status of anthropology. All of them are by-products of the same workshop in which my major work, *Woman's Evolution,* was fashioned; this collection can be regarded as a sequel and supplement to *Woman's Evolution.*

Evelyn Reed
November 1977

Primatology and Prejudice (1977)

Primatology is a new term for a special branch of science, the study of monkeys and apes. The study itself is not new; it has been going on to some extent ever since Darwin brought forward the primate origins of humankind. The word *primatology* came into use in the early 1960s with a sudden sharp rise in the collection of data on the habits of these animals.

This spurt of interest took off in two main directions, with some overlapping between them. On the one hand, it was hoped that experimental research on laboratory specimens would yield useful information for biochemistry, medicine, physiology, psychology, etc. On the other hand, field studies of primates in the wild were expected to shed light on human behavior through a greater understanding of primate life. Such observations would be of great service to anthropologists and others concerned with what is usually called the "science of mankind," which must necessarily begin with the transition from ape to human.

Unfortunately, primatology was born amidst great difficulties. Steady encroachment by humans on their former habitats had reduced the number of "wild" animals in the world; the free-roamers had been pushed onto ranges or reservations. To one degree or another they had been altered, some by crowding, others by contact with humans. Along with these practical problems was the serious theoretical deterioration which affected the science of anthropology and cast its shadow over the newly emerging primatology.

The majority of twentieth-century anthropologists, hostile to the evolutionary method of the nineteenth-century founders of the science, had long since replaced any comprehensive theoretical approach to their discipline with descriptive field studies. Many primatologists followed the same narrow empirical course, side-stepping general theory and restricting themselves to particular studies of different species of primates.

Empirical studies are essential for the development of any

science; they furnish the evidence required to sustain, verify, or invalidate a thesis. However, these cannot become a substitute for a systematic theoretical outlook. As Stuart A. Altmann points out, "There is a certain danger, in the present rash of primate field work, for empirical work to progress unreasonably faster than the theoretical." He adds, "The moral is obvious: the empirical and theoretical work should proceed in parallel" (*Social Communication Among Primates*, pp. 375-76).

The evolutionary method is fundamental to any theoretical exposition of the emergence of humans from primates. Primatologists who avoid this approach can easily give the misleading impression that contemporary primates are equivalent to ancient primates. A million years ago, when only a tiny number of humans existed, the world was preponderantly populated by wild animals, including large numbers of primates. Today that ratio has been drastically reversed. There are nearly four billion humans in the world. Wild animals, however, are so shrunken in numbers and hemmed in by civilization that a great many are in the category of endangered species. It would be unscientific to equate the behavior of the few surviving primate species today with their own animal ancestors, much less with humans.

Some scholars have warned against these pitfalls. In a joint paper, S. L. Washburn and D. A. Hamburg point out, "A central problem in the study of the evolution of behavior is that contemporary monkeys and apes are not the equivalents of human ancestors." They further caution against making field studies a replacement for the evolutionary approach or substituting studies of animals for the "direct study of man." In a rebuke to certain of their colleagues they write, "we echo Simpson's 1964 statement, '100 years without Darwin are enough'" ("Aggressive Behavior in Old World Monkeys and Apes," in *Primates*, p. 459).

By contrast, let us examine the theoretical fundamentals laid down by earlier scholars, even before primatology received its name.

Theoretical Fundamentals

By virtue of their evolutionary approach, naturalists, anatomists, and biologists were able to establish the sequence of stages in animal life, leading from the fish—the earliest vertebrate species—up to later and higher mammals, with the primates standing at the peak of the purely animal line of development. Within the primate order, anthropoids, or apes, stand higher than monkeys. At the same time, the four existing species of apes—the gorilla, chimpanzee, orangutan, and gibbon—are only remote cousins to the ancestral apes that gave rise to humans.

The pioneers in primatology also pinpointed the key biological organs and environmental factors that had developed over many millions of years of animal evolution to show how and why humans could not have emerged from any species lower than the higher apes.

Foremost among the biological prerequisites was the freed hand—that indispensable organ required for the tool-making human. While four-footed mammals "nose" their way around, the primates developed upright posture and a separation between the functions of the hands and feet. Instead of seizing food on the ground with their teeth, primates can pluck edibles and convey the food to the mouth. They can manipulate objects, such as sticks and stones, in a manner suggestive of tool-use. The impressions gained through the use of the hand stream into the brain, making these animals superior in intelligence to all lower species. The British anatomist F. Wood Jones has spelled out in detail how the activities of the hand led to the superior brain of the ape (*Arboreal Man*).

W. E. LeGros Clark, of the British Museum of Natural History, traces the evolution of primates from the tree shrew (the earliest form) through monkeys such as the lemur and tarsier to the macaque and up to the gibbon and chimpanzee. At the same time he cautions his readers against taking "a linear sequence of evolution," as though contemporary primates are the same as the ancestral stock or that humans are in the direct line of any of the existing species. He writes, "It must not be inferred, of course, that Man was actually derived from a chimpanzee ancestor, or that a monkey ever developed from the sort of lemur which exists today" (*History of the Primates,* p. 47).

Along with the hand and brain, another key factor contributing to the biological superiority of the primates was the extended period of mother-care provided for the offspring as contrasted with lower mammalian species. Rats or rabbits give birth to litters of from four to twelve offspring at a time. They reach sexual maturity in six to eight months and are full-grown in a year. A baboon monkey bears one offspring at a time, which reaches sexual maturity at three and is full-grown at five. At the

An adult female tarsier (left). The development of the hand through arboreal life goes back to this species, which preceded the monkeys and apes. Female langur (monkey) and offspring (right). Maternal care is highly developed among the primates, especially the higher apes.

top of the scale are the higher apes; a gorilla mother gives birth to one child, which reaches sexual maturity at about eight to ten and is full-grown at twelve to fifteen. This slow rate of development is closer to humans than to monkeys; a human child reaches sexual maturity at thirteen and is full-grown at twenty to twenty-five.

This slow maturation of infants occurs only where mother-care is prolonged, as it is among the higher apes. This is of paramount importance in the development of their advanced traits. In the lower animals, where mother-care is of short duration, the offspring are obliged to mature rapidly to become self-supporting. Among primates, which mature much more slowly, the young animals can learn from imitation and experience, modify their behavior patterns, and acquire greater reasoning abilities and intelligence.

Robert Briffault has analyzed in detail the prime importance of prolonged mother-care in developing these traits. (*The Mothers,* vol. I.) More recently, Washburn and Hamburg make the same point. They write that monkeys and apes "mature slowly and there is strong reason to suppose that the main function of this period of protected youth is to allow learning and hence adaptation to a wide variety of local situations" ("Aggressive Behavior in Old World Monkeys and Apes," in *Primates*, p. 464).

But there is another side to this prolonged period of mother-care—its effect upon the females themselves. The more extensive functions of the females in providing for and protecting their infants, together with the longer periods in which they exercise these functions, make the females the more intelligent, capable, and resourceful sex. This aspect has also been dealt with by Briffault. It may explain why females are so often selected for intelligence tests and experiments. As one writer complains, "all the intensively studied individual chimps, including those in the language experiments, are females. It is time that male chimps demand equal treatment" (Joseph Church in a review of Ann J. Premack's book, *Why Chimps Can Read, New York Times Book Review,* April 11, 1976).

Tracing the line of continuity from lower to higher forms of animal life enabled the evolutionary biologists to show how and why humans ascended from a branch of the anthropoid species and no other. However, to appraise human life it is necessary to go beyond the *continuity* of animal evolution as such to the point at which a definitive *discontinuity* occurred—when a jump was made into a totally new kind of evolution—human social evolution.

Many scholars slur over this vital distinction between biological and social evolution. For example Jane B. Lancaster writes that "mankind has evolved and expanded in accordance with the same major evolutionary processes as have other species of animal life" (*Primate Behavior and the Emergence of Human Culture,* p. 1). However this applies to human evolution only up to that point where the ape became hominid.

George Gaylord Simpson has emphasized that unlike all animal species, which have evolved through only one kind of evolution (organic, or natural evolution), humans have evolved through a wholly new kind—social evolution (*The Meaning of Evolution*). Moreover, the new social evolution increasingly displaced the old biological evolution, to the point that today humans have lost virtually all their ancestral animal patterns of behavior and instincts. These have been replaced with their own socially conditioned reactions.

In other words, to understand what a human being is, it is not sufficient to analyze the biological *preconditions* required for humanization. It is also necessary to uncover the special and, indeed, unique *conditions* upon which human life rests, and without which it cannot survive. This problem was first clarified by Frederick Engels in his essay, "The Part Played by Labor in the Transition from Ape to Man" (*Origin of the Family, Private Property, and the State*). Our branch of the higher apes, equipped with hands, began to make and use tools in systematic labor activities. Production and reproduction of the necessities of life—which no other animals are capable of—became the prime conditions for human survival and progress. This remains so to the present day.

Tool-making and labor activities, therefore, represent the starting point for differentiating between humans and animals since these activities represent the foundation for social life. As Kenneth P. Oakley puts it, "Man is a social animal, distinguished by 'culture': by the ability to make tools and communicate ideas" (*Man the Tool-Maker,* p. 1). More recently, John Napier, emphasizing the significance of this title, wrote, "Probably the most generally accepted definition of man at the present time is that of man-the-toolmaker" ("The Locomotor Functions of Hominids," in *Classification and Human Evolution,* p. 178).

Even the biological structure of humans changed under the

impact of their laboring activities. They lost their hairy coats, acquired full upright posture, and developed hands with a completely opposable thumb. Behavior patterns changed too. Humans were obliged to curb and suppress their former animal individualism and competitiveness, replacing these traits with the social and cultural rules required for the establishment of human life.

The most important long-range result of the emergence of humans is that they alone have been able to transcend the barriers that keep animals within biologically circumscribed limits. It took scores of millions of years for the fish to evolve into the mammal and more millions of years to reach the higher-ape species. Yet at the end of this billion-year process, all animals, including the more flexible apes, remained chained to their biological limitations. Only one branch, our own, was able to break these fetters and acquire the *unlimited* possibilities inherent in the human capacity for labor, for changing themselves, and for developing new capabilities as they secure ever greater mastery over nature.

This qualitiative distinction was emphasized in a review in the April 1977 *Scientific American* of a recent book on the Kalahari hunter-gatherers, edited by Richard B. Lee and Irven DeVore, which made the following point: "The food and water of the /Gwi are won in those [barren] months chiefly by the digging stick. Hard work retrieves from the cool subsoil two species of tubers that provide food and a bitter tuber that yields water. The clever baboons that dwell in most of the desert cannot live hereabouts, because they are not masters of the digging stick; only the much cleverer human beings can survive."

Gordon Childe hypothesizes why only one branch of the higher apes was propelled into this revolutionary change from animal to human. The coming of the Ice Age about a million years ago produced catastrophic changes in climate affecting virtually all species on earth. The struggle for survival took on gigantic proportions. Some species were completely wiped out. Our branch of the anthropoids may have been too far advanced biologically for any further animal adaptations, because it was at this point in time that the first tool-using hominids made their appearance on earth. As Childe writes, "The most curious of all the species emerging was, however, Man himself" (*What Happened in History,* p. 29).

The same point is made by William Howells, of the American

Museum of Natural History, who writes, "It is extraordinary that the sudden, severe Age of Ice, a mere pinpoint in time, should have coincided with the very period, also short, when man at last was rapidly becoming what he is today" (*Mankind So Far,* p. 113). Our ancestors met the severe challenge of nature by making tools and working for a living, thereby passing over from primate existence to a human mode of life.

These theoretical fundamentals were established by various scholars before primatology came into existence as a distinct science. This raises the question: to what extent have these guidelines furnished the background for the interpretations made by primatologists of the animal behavior they are studying? The answer is that while some have adhered to the genuinely scientific approach, many have not. The latter appear to be heavily influenced by the antievolutionary attitudes that dominate anthropology today. This leads them to cater to two erroneous themes: humans are little more than primates; and females have always been the inferior sex, dominated by males. Let us examine both of these propositions.

Are Primates Tool-Makers and Meat-Eaters?

For a time scientists tried to find the "missing link" between apes and humans by comparing fossil bones and skulls. When this purely anatomical test failed because the differences during the transition were so minor, they made tools the criterion by which to distinguish between the two. If human-made artifacts were found in the same deposits as the bones and skulls, this provided evidence of "Man the Tool-Maker."

Engels, however, did not leave the matter at this point. He explained that tools were the instruments of labor activities; therefore in its most comprehensive sense, labor activities represent the point of departure from animality to humanity. Stephen Jay Gould, a Harvard geologist, recently called this proposition the "missing link." At the same time he admitted that "few scientists were ready to recognize the 'missing link' when we found it" ("Posture Maketh the Man," in *Natural History,* November 1975).

The labor theory of social origins was ignored by those who were determined to blur the dividing line between humans and primates. Irritated perhaps by the important role this discovery gave to "lowly" labor activities, some investigators began to undermine even tool-making as the distinctive acquisition of humans. K. R. L. Hall writes that "the criterion of tool-using is no longer used by anthropologists to signalize a supposedly critical stage in the transition of ape to human" ("Tool-Using Performances as Indicators of Behavioral Adaptability," in *Primates,* p. 144). He attributes this to a "tendency to overestimate the significance of such performances . . . largely because of the urge to discover equivalences to stages in human evolution" (ibid., p. 146).

Thus, the denial by academic anthropologists that humans had passed through a sequence of stages in social evolution was extended to primatology. The importance of tool-using and tool-making began to be downgraded under pressure to conform to this antievolutionary doctrine. This retreat has gone further as some primatologists have begun to overestimate the ability of primates to manipulate sticks and stones and have equated them with human tool-makers. In both instances, these represent attempts to liquidate the qualitative distinction between humans and animals.

Jane van Lawick-Goodall became most famous in this attempt to elevate primates to tool-makers. She began her studies of chimpanzees in the Gombe Stream Chimpanzee Reserve in Tanganyika in the early 1960s. By the end of the decade she made headlines with her proposition that not only man but primates were users and makers of tools. This wasn't all. She also suggested that meat-eating, like tool-making, was a trait shared by humans and primates alike. This was sensational because scientists had hitherto regarded both tool-making and meat-eating as the two major acquisitions of humans *after* their departure from animality. As E. Adamson Hoebel wrote, the carnivorous diet "more than any other single trait distinguishes [man] from his vegetarian anthropoidal relatives" (*Man in the Primitive World,* p. 102). Goodall was expressing a contrary view.

According to Emily Hahn, both of Goodall's propositions "came as a surprise to the world of anthropology" (*On the Side of the Apes,* p. 154). Whether greeted with surprise or delight, Goodall's views showed the powerful influence of the antievolutionary anthropologists upon the nascent science of primatology.

This point was made explicit by Maggie Scarf in an article-interview published in the February 18, 1973, *New York Times Magazine.* "What is the explanation for Goodall's widespread appeal?" she asks. Her answer is that "Goodall's work has led many scientists to reassess the 'great gulf' which has been believed to separate the animal and human worlds." Scarf writes that according to Irven DeVore, Goodall's data "strongly suggests that the gradation between what might have been our chimp-like ancestors and a very early hominid or true human represents a small step rather than a great leap." Yet it was precisely that "small step" taken at a critical juncture of evolution which became the starting point of the vast—and still growing—gulf separating humans from animals.

Goodall, who acknowledges that she undertook her studies with a mind "uncluttered" by theory, confines herself to descriptive observations. Thus, she found that some chimpanzees could defoliate a twig, insert it into a crevice to get at termites or ants, and then convey the insects on the stick to the mouth. This "stripping off of the leaves," she says, shows "a wild animal not merely *using* an object as a tool, but actually modifying an object and thus showing the crude beginnings of tool*making* (*In the Shadow of Man,* pp. 6, 37). Other performances include the crumpling of a leaf to make a "natural bowl" to lift water to the mouth.

It has long been recognized that apes, with their flexible hands, are capable of manipulating various objects, and that in captivity or under human influence they can be very clever at these practices. Frederick Tilney, citing observations of chimpanzees made by Wolfgang Köhler during World War I, writes that those chimps usually ate insects by rolling their tongues over them, but sometimes they "used straws and twigs as we use spoons." They also used straws to suck water up to their mouths. However, as Tilney points out, "the handling of everyday objects by the chimpanzee comes almost entirely in the nature of play" and not through any necessity to use these objects as tools (*The Master of Destiny,* p. 190).

These playful performances cannot be characterized as tool-use or tool-making in the proper sense. Ape survival does not depend either upon eating insects or using twigs to get at them. Ape survival, past and present, depends upon a sufficiency of fruits and vegetation and upon the hands with which to grasp the food and convey it to their mouths. Eating insects and defoliating twigs is only incidental and episodic in the life of an ape. By contrast, under the lash of drastic environmental changes, our progenitors were forced to make and use tools—to labor—in order to survive. Labor activities remain the elementary basis of survival for humans to the present day.

Goodall's observations of apes provide useful clues as to how humans became tool-users. But they do not alter the fact that no other primates, past or present, can be regarded as identical with the species that alone bridged the gulf between the animal and the human. As Jane Lancaster, an admirer of Goodall's work, observes, "The evolution of skilled tool-using marks a major change from the kind of tool-use that is incidental to the life of a chimpanzee to the kind that is absolutely essential for survival of the human individual" (*Primate Behavior and the Emergence of Human Culture,* p. 53).

The same holds true for the human diet of regular meat-eating. Most animal species are rigidly restricted not only to specific habitats but to particular sets of foods, and they cannot survive on any other. Carnivores cannot subsist on grass or other vegetation and ruminants cannot eat meat. Even under human influence, they cannot change their diets. Although primates are vegetarian animals, they are more adaptive than the lower species. In captivity or through changes in their environment, some can assimilate meat. Here again we have a clue about how our ancestral branch became the first omnivorous species—human hunters and meat-eaters. But this doesn't alter the fact that primates are vegetarian animals, most of them eating no meat at all.

Even among those which are most adaptive in this respect, meat constitutes only a tiny morsel of their diet. The data on this is unambiguous. Robert M. Yerkes wrote of chimpanzees as naturally and primarily vegetarian—their staple diet consisting of fruits, seeds, blossoms, leaves, shoots, and the bark of many African plants. They also eat eggs and small organisms. In captivity, however, an "occasional specimen may become omnivorous or carnivorous" (*Chimpanzees, A Laboratory Colony,* p. 222).

More recently George B. Schaller wrote of the gorilla, "I never saw gorillas eat animal matter in the wild—no birds' eggs, insects, mice, or other creatures—even though they had the opportunity to do so on occasion. . . . In captivity, however, gorillas readily eat meat" (*Year of the Gorilla,* p. 180). Emily Hahn remarks about her pet, "Chimpo ate meat whenever I did" (*On the Side of the Apes,* p. 154). According to Biruté Galdikas-Brindamour, "wild orangutans have never been known to eat meat," although "they have been observed munching insects and birds' eggs" (*National Geographic,* October 1975, p. 468).

Even Jane Goodall, who plays up the meat-eating proclivities of chimpanzees, gives statistics that show these animals to be preponderantly vegetarian. "More than 90 different species of tree and plant used by the Gombe Stream chimpanzees for food have already been identified. They have been eating over 50 types of fruit and over 30 types of leaf and leaf bud. They also eat some blossoms, seeds, barks, and piths. Sometimes they lick resin from tree trunks or chew on wads of dead wood fiber" (*In The Shadow of Man,* p. 281).

In fact, from her account, meat-eating may be no more than what Goodall calls a "craze," that is, an experimental and haphazard chewing and tasting which includes many other things. She describes one time when "little groups of chimps had sat around tearing up and chewing chair seats, flaps of tents—even Kris's camp bed had been destroyed. . . . the back of one of Hassan's homemade cupboards had gone, and the leg of a wooden chair" (*In the Shadow of Man,* p. 92).

Moreover, in establishing her rapport with the animals she was studying, Goodall did not entice them with meat but with their favorite fruit—bananas. Her feeding station consisted of endless boxes of bananas. "An adult male, if he has the chance, will eat fifty or more bananas at a sitting," she writes (ibid., p. 89).

Baboons furnish the favorite example of "meat-eating" primates. These animals are not classified as higher apes, and are more specialized for terrestrial life than other monkey species. Baboons have heads that resemble those of dogs, and they go about more like four-footed animals on hands and feet. The males are equipped with long and dangerous canine teeth and are highly competitive and aggressive.

According to Ivan T. Sanderson, "the life of baboons is a hard one, for they live in territory that, although often rich in food for ungulates, great cats and carrion feeders, is poor in smaller fare such as that which suits omnivorous primates, like bulbs and succulent roots, insects, fallen fruits and so forth." They are, he says, "perpetually at war with the carnivores" (*The Monkey Kingdom,* p. 136). It is not surprising, therefore, that these animals, under certain circumstances, have learned how to eat meat.

Even so, meat is only an incidental portion of their diet. According to Washburn and DeVore, "baboons are almost entirely vegetarian. They eat a wide variety of fruits, buds, leaves, grasses and roots. Eggs, nesting birds, some insects and larvae are also eaten." Although "a small live animal in the grass will also be killed and eaten" their normal diet "is overwhelmingly vegetarian, and animal food accounts for less than 1 per cent" ("Social Behavior of Baboons and Early Man" in *Social Life of Early Man,* p. 94). In short, baboons are not equivalent to human hunters and meat-eaters.

Since hunting by humans involves using and making weapons, some writers exaggerate the ability of primates to pick up and hurl sticks and other objects as though they, too, can use and make weapons. This is no more valid than the claim that these animals can make and use tools. Tilney writes, "Any angry monkey may in its rage grasp and hurl an object . . . but there is usually no accuracy in its aim or intention in its act other than an expression of irritated feelings. None of the monkeys has ever been known to use a stick or a club in attacking others or defending itself" (*The Master of Destiny,* p. 157).

More recently, K. R. L. Hall remarks that "even the clear statements of 'purposiveness' in such acts, as in Wallace's (1902) report on the orangutan, and Carpenter's (1934, 1935, 1940) records for the howler, red spider monkey, and gibbon, must be used comparatively with the same sort of caution that one should bring to bear upon relative estimates of tool-using skills in laboratory primates" ("Some Problems in the Analysis and Comparison of Monkey and Ape Behavior," in *Classification and Human Evolution,* p. 293).

From all the evidence, then, the question, "Are primates toolmakers and meat-eaters?" must be answered in the negative. They are merely the most adaptive or preadapted animals, whose habits furnish clues to the reconstruction of the transition from our branch of the higher apes into the tool-making, meat-eating humans.

Equating humans with primates by overemphasizing certain similarities while underplaying the vast differences is unscientific. One of the unfortunate results of this error has been the misapplication of the term "society" to animal life.

Do Primate "Societies" Exist?

The attack upon the evolutionary approach to human life and development has itself gone through an evolution. In the first stage, after Darwin, the primate origin of humankind was categorically denied. It was asserted that humans had always been humans and never animals. As the scientific evidence to the contrary piled up, this assertion was displaced by its opposite: humans have always been essentially animals, comparable to monkeys and apes.

This disavowal of the uniqueness of humanity sometimes takes subtle and indirect forms. One is in the realm of terminology. Primates were formerly referred to as animals, distinct from "man," the human. Today the term "nonhuman primate" is widely used, inferring, without explicitly saying so, that there is a "human primate" not essentially different from the "nonhuman primate." Also, the term "society," which belongs exclusively to human organization, has been bestowed upon an endless range of species, from primates down to ants and termites. So far oysters, worms, protoplasm, and genes have been excluded, but perhaps some biological determinists will soon elevate these clusters into societies too.

To be sure, every growing branch of science confronts a problem with terminology—finding words for new phenomena. The automobile, for example, was called the "horseless carriage" before it acquired a name of its own. Similarly, the term "social behavior" was used in connection with the study of animal interactions and aggregate behavior. Thus, not only are primate troops called "societies" but so are lower mammalian species such as lions, wolves, elephants, etc. These "societies" should more properly be referred to as primate troops, elephant herds, wolf packs, lion prides, etc.

According to Gordon Childe's definition, society is "a cooperative organization for producing the means to satisfy its needs, for reproducing itself, and for producing new needs" (*What Happened in History,* p. 17). While animals, like humans, reproduce themselves, no animal species, including the higher apes, have ever produced for their needs; much less have they produced new needs. From the time of the first digging stick and fist axe, these new needs have accumulated at an ever accelerating pace, spurred on by progress in productive know-how. Only humans can produce the necessities of life as well as produce new needs and the means for their satisfaction.

Production, in turn, cannot exist except in an organized social arena in which all the adult members coordinate their efforts and collectively provide for themselves and for the young and old members of the group. As Childe points out, the experiences gained in productive activities are pooled, and the techniques are passed on through precept and language to new generations of producers, who gradually improve upon the work done by their predecessors. Language, traditions, customs, and culture grow up out of this social organization, which is based upon production. In its essential meaning, therefore, no animal aggregate qualifies for the term "society."

This has not halted the numerous attempts to erase the gulf between animal existence and human society. One instance can be found in the realm of communication. All animals do have some form of communication, whether by contact, sight, smell, or sound. Because of this, some biological determinists contend that primates, like humans, have the ability to talk and use language. They take certain experiments performed with laboratory or pet animals, exaggerate their results, and declare that primates can talk, or learn to talk, like humans.

At a recent Academy of Science conference held in New York, the authoritative S. L. Washburn stated flatly that "no chimpanzee can speak, and they can't be taught to speak" (*New York Times,* September 25, 1976). The same point is made by Joseph Church, a psychology professor, reviewing Ann J. Premack's book *Why Chimps Can Read.* Although, he says, they understand some human language, so does the family dog. Thus, "chimps, whether reared in the wild, in the laboratory, or in the home, seem to have severely limited communication skills. No amount of exposure to human language, with or without special training, has led any known chimp to try to talk" (*New York Times Book Review,* April 11, 1976).

Jane Goodall likewise writes that "the chimpanzee has not developed the power of speech. Even the most intensive efforts to teach young chimps to talk have met with virtually no success. Verbal language represents a truly gigantic stride forward in man's evolution" (*In the Shadow of Man,* p. 248). All this confirms what certain evolutionary scientists have long contended, that speech grew up out of the cooperative productive efforts of humans banded together in a social organization.

Primates are not capable of such sustained cooperative efforts. Frederick Tilney cites a graphic example of a group of chimpanzees obliged to build a tower of boxes to get at a basket of food hung out of their reach. They could not learn "the advantages of a mutual aid society" and were unable to develop "an efficient labour organization," he writes. Separately they could manipulate bamboo sticks and boxes, but their building operations were constantly impeded by their individualism. Getting at the food was "a matter of the utmost selfish concern to each chimpanzee." As a consequence "of this highly individualistic competition . . . the tower would sometimes tumble over" (*The Master of Destiny*, pp. 199-200).

The cooperation that does exist in the animal world is largely restricted to the females and their young and stems from the females' reproductive functions. Females provide for and protect their offspring; among primates mother-care endures for a longer period than with lower animals. Females also form affective bonds with other females. Shirley C. Strum writes about the Kekopey baboons: "I observed only two types of baboon friendships, those between females, and those between males and females—never between adult males." She adds, "friendships between males and females seemed neither so intense nor so lasting as between females." She writes that "among baboons, as among most monkeys and apes, paternity is neither known nor recognized. A family comprises an adult female with all her offspring, including large sons not fully mature" ("Life with the Pumphouse Gang," in *National Geographic*, May 1975, p. 680).

Many primatologists have noted that the "stable unit" of a primate troop is composed of mother and offspring; males are peripheral and often solitary individuals. A number of such mothers and their combined offspring comprise what is often called the "central core" of a troop, and it is within this core that affective sentiments and a certain degree of cooperation can be found. Although males start roaming once they reach adulthood, the females remain together, often for life.

It is among the females and young, therefore, that modifications take place in animal behavior, which in the animal world is generally individualistic and competitive. This is a clue as to why the females in our branch of the higher apes led the way over the bridge from animality to humanity. From biological mothers in the primate world, they proved capable of becoming social mothers in the human world by founding the maternal clan system of social organization. These "motherhoods" provided the models for the cooperative "brotherhoods" by which males were assimilated into the social structure.

Since this reconstruction of the earliest society is denied or rejected by most anthropologists today, it has yet to be considered or accepted by primatologists. Many of them accept the male-biased proposition that all societies, past and present, have been dominated by men, with women occupying a submissive and subordinate status. This thesis is projected back into the animal world, where the primate male is portrayed as the sultan of a harem, or patriarch of a family.

One of the arguments put forward to buttress this thesis is the "sexual dimorphism" that exists in animal life—that is, the difference in size between males and females. In fact, this sexual dimorphism varies with different species of primates and varies even within the same species. Adolph H. Schultz writes that

> sex differences in average body weight vary to an amazing degree among primates and this even in closely related forms, such as gorillas and chimpanzees or different species of macaques. While in some species the males can attain weights of more than twice those of females, in other species the females are on an average somewhat heavier. . . . In gibbons, siamangs and chimpanzees one frequently finds some males smaller than some females of the same local populations. . . . Even among adult orangutans with their great *average* sex difference in size, the writer has seen evident runts and giants in both sexes ["Age Changes, Sex Differences, and Variability as Factors in the Classification of Primates," in *Classification and Human Evolution,* p. 93].

Primate sexual dimorphism began when some primates left the shelter of tree life to become more or less terrestrial animals. There is little difference in size between male and female gibbons, which are arboreal animals. Baboons, however, inhabit rocky, open country and are surrounded by the large carnivores. As Irven DeVore remarks, "Life on the ground exposes a species to far more predators than does life in the trees" ("A Comparison of the Ecology and Behavior of Monkeys and Apes," in *Classification and Human Evolution,* p. 313). Males in this species are usually larger than females, and they also have long, sharp, canine teeth.

Nevertheless, it is incorrect to view male animals as protectors of their "families" or defenders of their "societies" as humans are. All animals defend themselves either through fight or flight. As Washburn and Hamburg put it, "individual animals must be able to make the appropriate decisions and fight or flee" ("Aggressive Behavior in Old World Monkeys and Apes," in *Primates,* p. 463). Females will fight to protect their young but prefer to flee. Males will often turn and fight. But they do so as individuals and not as protectors of their "families." Females are extremely adept at concealing themselves and their offspring; they do not depend upon males for their protection. Only humans can organize guards, soldiers, and other defense bodies which are found in social life.

Another misconception is the notion that males are usually the leaders of primate troops and sound alarms, or give other signals at the sight of a predator, then conduct the group to safety. In fact, *any* animal that senses danger may utter the cry of alarm that alerts the others. Moreover, adult males, peripheral to the central core of females and young, are separated from it whether they are in the front, the rear, or to the side of the troop. Recognizing this, K. R. L. Hall writes that "the distance of the male away from the group is perfectly compatible with his defensive function as a watcher and diverter rather than as an aggressive protector" ("Behavior and Ecology of the Wild Patas Monkeys in Uganda," in *Primates,* p. 110).

This confirms the fact that even if a male is ahead of the troop, this does not make him its "leader." Carl B. Koford notes with respect to a troop of rhesus monkeys, "the presence of the leader himself is not always necessary, for on a few occasions a leader was held captive and his band had the same precedence as before" ("Group Relations in an Island Colony of Rhesus Monkeys," in *Primate Social Behavior,* p. 149). According to some primatologists, old females are the actual leaders of the troop whether or not some males are in a peripheral position ahead of them.

Moreover, primate protection can come not only from their own members sounding an alarm but even from alien species which also fear the large carnivores. Washburn and DeVore point out that baboons will commonly walk among noncarnivores like antelopes, gazelles, etc., with whom they establish a kind of indirect "interdependence." "The ungulates have a keen sense of smell, and the baboons have keen eyesight. If [the baboons] see predators, they utter warning barks that alert not only other baboons but also any other animals that may be in the vicinity. Similarly, a warning bark by a bushbuck or an impala will put a baboon troop to flight. A mixed herd of impalas and baboons is almost impossible to take by surprise," they write. They also point out that a baboon troop's "ultimate safety" is to flee to the trees ("The Social Life of Baboons," in *Primate Social Behavior,* p. 102).

From the evidence, then, there are no primate "societies."

However gregarious, intelligent, and beguiling these creatures are, they remain animals. They cannot form cooperative labor units, develop language, or acquire any of the cultural traits and institutions which came into existence with the human species alone. These strictly human creations are the fundamental elements that go into making the structure of "society."

Despite this, it will probably take some time before the term "primate society" is discarded. The use of this term makes it easier to uphold the myth that primate males are the dominant sex, the leaders of their "societies," just as men are in modern society. By the same token, it is easier to disseminate the companion myth that the female inferiority which characterizes class society today goes all the way back to the primates.

Are Females the Inferior Sex?

The thesis that males are socially superior to females rests upon two biological traits: the males are often larger and more muscular, and the females bear the offspring. The females are then portrayed as helpless and dependent upon the brawny, brainy males for sustenance and protection—in the animal as well as the modern human world. But after the evidence piled up that all animals forage on their own and that males do not provide food for females, proof of the thesis narrowed down to male physical superiority. Males can "dominate" females.

The anthropologist Ralph Linton puts it, the males of *Homo sapiens* are "on the average, larger and heavier than the females and able to dominate them physically. Whether the feminists like it or not, the average man can thrash the average woman." Combined with this endearing trait is the "continuous sexual activity" of the males, who "are actively interested in all females and try to collect and hold as many of them as possible." Jealous of rival males, they "restrict the attentions of their female partners to themselves." He concludes that the dominated females "are in no position to do anything about it. The double standard is probably as old as the primate order" ("Selections from 'The Study of Man,'" in *Sex Differences*, p. 182.)

Apart from his fictional view of the matter, Linton's testiness is quite understandable. Nine years before Linton wrote his book in 1936, Robert Briffault had produced his monumental work, *The Mothers,* which presented his matriarchal theory of social origins. He showed that maternal functions, far from being a disadvantage, were one of the most essential biological factors elevating our branch of the higher apes into labor activities and social life.

Antimatriarchists, including Linton, closed ranks against Briffault and his supporters. That male muscles do not represent a major force in social history and that women have not always been dominated by men was heresy to them. As Linton put it:

> The physical superiority of the human male has had a much greater effect on the development of social institutions than we usually realize. . . . Among the sub-human primates the male can dominate a group of females because these females are unable to organize among themselves. He can deal with them in detail [ibid., pp. 184, 190].

Contrary to this crude male-biased thesis, primate males do not thrash females, do not possess "harems," and do not have control over the sexual activities of the females. Females have the advantage because they outnumber males in primate troops and are capable of cooperating with one another. If a male gets out of hand, the females can band together to teach him a lesson.

The "harem" theory of male dominance is based upon the fact that in primate troops females outnumber males, often by two-to-one, or more. This gives them an aggregate power that is superior to the muscles of any individual male. In baboon troops, according to Washburn and DeVore, there are twice as many females as males; Shirley Strum gives a figure of three times as many females as males. C. L. Carpenter gives examples among other species. One macaque group consisted of two adult males to six adult females; another had four males to ten females. A rhesus group contained six males and thirty-two females; another had only seven males to seventy-eight females. A spider monkey group had eight males and fifteen females. For howler monkeys there were twice as many females as males. ("Societies of Monkeys and Apes," in *Primate Social Behavior,* pp. 29, 31, 32, 40.) In all cases infants and juveniles swell the female portion of the troop, which moves together as a separate pack within the troop.

From the male-biased point of view, this paucity of males gives each one a large number of females to "dominate." The troop is then portrayed as a group of "harems," each under the overlordship of a patriarchal master. According to Carpenter, "A despotic male baboon may continuously and persistently possess five or six females," and among langurs "a single despotic male may dominate an entire group of ten or twelve females and their young, while all other adult males are excluded from association in the group." To him these are vivid examples of "autocratic male dominance" which affect the "grouping pattern" (ibid., p. 41).

In fact, the behavior that Carpenter calls "despotic" is no more than the aggressive behavior of one male seeking to dominate other males and win access for himself to the female group. That is why there are so few adult males in a primate troop; the dominant ones have either ousted or subordinated the others. But this individualistic aggression makes even the dominant male vulnerable to other males, and there are very few who enjoy any extended tenure as "despots." Shifts from one dominant male to another constantly take place.

Instances have occurred where males have threatened, injured, or killed females and their young. But these situations occurred with captive animals or with those too crowded together to enable them to live normal lives. The best known example is the London Zoo experiment of 1925-30 which resulted in the catastrophe described by Solly Zuckerman. A small number of female baboons were placed in an all-male compound. The females were outnumbered by the males, and there were no retreats provided for them. The fights which ensued resulted in the virtual extermination of the females and young. (See *Woman's Evolution,* pp. 62-67.)

A more recent example given by K. R. L. Hall tells of baboons in captivity at the Bloemfontein Zoo, where, through the introduction of an "alien" pair, fights broke out and most of the animals were killed, or died of their injuries. He concludes that "it is all too well known that, in the unnatural restriction of physical and social space of the usual captivity conditions lethal aggressiveness may occur" ("Aggression in Monkey and Ape Societies," in *Primates,* p. 155).

Such situations are not known to occur in the wild, and it is therefore misleading to view these examples as indicative of normal baboon behavior. As Irven DeVore, citing other authorities, writes with respect to Zuckerman's account, "behavioral observations on confined animals can be very misleading, however, and no field study of baboons has found that sexual jealousy or fighting is frequent in free-ranging troops" ("A

Comparison of the Ecology and Behavior of Monkeys and Apes," in *Classification and Human Evolution,* p. 313).

In the wild, females control their own sexual activities, and males adapt themselves to the estrus periods of the females. During the extended periods when females are caring for their offspring they do not seek sexual congress, and males do not approach them. Phyllis C. Jay writes, when a female "is not in estrus, the adult males show no sexual interest in her. She is the sole initiator of sexual activity, and she is not mounted unless she solicits the attention of the male" ("The Female Primate," in *The Potential of Woman,* p. 6).

When she is sexually receptive, the female is even more vigorous than the male, and far from restricting her intercourse to a solitary male, she mates with many males in the vicinity. Even Carpenter, who views the male as a patriarchal despot, admits that "the females in natural groupings of primates are usually the aggressors and initiators of sexual responses. . . . During estrus, a female's capacity for copulation greatly exceeds that of any one male" ("Societies of Monkeys and Apes," in *Primate Social Behavior,* pp. 44-45). There are no reports of any male, even the dominant male, restricting "his" female's sexual attentions to himself. Most reports tell of males quietly awaiting their turn at access to the female.

Sexual pairing is incidental in the life of a female primate and usually short-lived. Washburn and DeVore write that a consort pair "may stay together for as little as an hour or for as long as several days." They "usually move to the edge of the troop. It is at this time that fighting may take place, if the dominance order is not clearly established among the males." However, "Normally there is no fighting over females, and a male, no matter how dominant, does not monopolize a female for long." They conclude, "there is nothing resembling a family or a harem among baboons. Much the same seems to be true of other species of monkey. Sexual behavior appears to contribute little to the cohesion of the troop" ("The Social Life of Baboons," in *Primate Social Behavior,* p. 108).

Sexuality contributes so little to the cohesion of the troop that the *segregation* of the sexes is far more pronounced than the brief pairing which occurs when a female comes into estrus. Due to these extended periods of sex segregation, the female primate bears few offspring in her lifetime. Biruté Galdikas-Brindamour writes that male and female orangutans are rarely even seen together. She says that males live a conspicuously solitary existence and that the cohesive group consists of the female and offspring. A female bears an offspring only once every four or five years. ("Orangutans, Indonesia's 'People of the Forest,'" in *National Geographic,* October 1975.)

There are similar reports of other species, including the more gregarious types. According to Goodall, there are relatively few births in a chimpanzee community; "mothers only have an infant about once every three and a half to five years" (*In the Shadow of Man,* p. 146). George Schaller writes that the female gorilla "gives birth only once every three or four years unless, of course, her child dies in infancy." He states that a close tie exists between mothers and their offspring, and that this bond persists for four years or more—long after the mother has ceased to protect and provide for them. (*Year of the Gorilla,* p. 134.)

In contrast with these more enduring relations between females and their young, the relations between adult males and females are ephemeral. Koford remarks that female rhesus monkeys "are rarely seen with a male, and yet they conceive" ("Group Relations in an Island Colony of Rhesus Monkeys," in *Primate Social Behavior,* p. 149).

Shirley Strum underscores the striking contrast between primate and human sexuality by observing that "female baboons spend more than 90 percent of their lives sexually nonreceptive, and it's not much less for males. It's very different from the human preoccupation with sex" ("Dr. Strum: Presenting Baboons," by Arnold Shapiro, *Westways,* March 1977, p. 39).

Thus, the "grouping pattern" of primates is not based upon the ephemeral sexual relations between males and females, but upon the more durable bonds of females and offspring. The animal "group" is not a family dominated by the father; it is a maternal brood in which the male is usually not even present. The segregation of the sexes is far more pronounced than their fleeting unions for sexual intercourse, which occur only in the mating season. Females represent the central core of the primate group.

Shirley Strum is one of the young primatologists who has questioned the male-biased versions of primate behavior. She writes:

> Adult males are viewed as the core of the troop, affording protection, asserting discipline, and providing cohesion through their leadership. The role of females is merely reproductive. . . . but from new

> information about primates, gathered worldwide, I had my doubts that even baboons could be so easily explained. I questioned that the adult males, a small percentage of the troop, could be responsible so totally for its social life ["Life with the Pumphouse Gang," in *National Geographic,* May 1975, p. 679].

Two years later, after further studies, she resolved this question in her own mind. She said:

> unfortunately many popular writers . . . have misapplied the data in order to sell books. They've used animal behavior as a justification for what the authors *like* to think about human behavior. . . . We can't use animal behavior to justify human foibles, and decide we're destined to behave with violence or force or male dominance because it's an irrepressible part of our "animal nature." These are misleading and dangerous assumptions [*Westways,* March 1977, p. 76].

Strum has a more objective approach to the study of primates. According to Arnold Shapiro, who interviewed her, she wants to see "what's shared and what's different" between primates and humans, and to distinguish between biological similarities and uniquely human attributes (ibid., p. 76).

The fact that men hold the leading positions in current society does not mean that this has always been the case. In the animal world, rudimentary forms of leadership are furnished not by the males but by the adult females. K. L. R. Hall, citing a personal communication from Sanderson (1957), writes

> an adult female almost always seemed to "lead" the groups he saw in West Africa. There is ample confirmation of this in the present study. On many occasions when a group had halted, as in a day-resting tree, it was one or more of the adult females who were the first to descend and move across the savannah. In such instances, the adult male might follow after most, or all, the rest of the group had gone after the leading females. The distances that separated the leading females from the adult male were sometimes considerable.

Nevertheless, says Hall, "it is an oversimplification of the social organization to say that the patas packs are '. . . bossed by old females'" ("Behaviour and Ecology of the Wild Patas Monkey in Uganda," in *Primates,* p. 112). If so, it is a simplification based on fact, not on prejudice.

Cooperation among female primates is sometimes referred to as "friendships," "coalitions," or "alliances." Some of these are permanent; others are temporary to repel a troublesome male. G. Gray Eaton writes that the adult female Japanese macaques

> form alliances with one another, a kind of behavior that is rarely observed in unrelated males. . . . If a male fights with a female, other females will usually come to her aid, whereas adult males rarely assist one another. . . . Female alliances therefore appear to be more important in regulating the cohesiveness of the macaque social order than sexual attraction between males and females, which in this species is seasonal and transitory ["The Social Order of Japanese Macaques," in *Scientific American,* October 1976, p. 102].

Phyllis Jay, writing about the langur, says that

> on six occasions adult males were the object of female alliances. The females rushed at the offending male who may have inadvertently frightened an infant. Generally the male made no attempt to threaten the females but instead ran a short distance. . . . Five instances were recorded of single adult females actually hitting or slapping a dominant adult male. . . . Usually the stimulus for the female attack is the frightening of her infant by the male. The adult male is usually taken unaware by the rushing onslaught of the female and no male was ever seen to strike the female in return ["The Indian Langur Monkey," in *Primate Social Behavior,* p. 121].

Jane Beckman Lancaster writes about vervet monkeys: "I often saw females form coalitions against the top three males when they monopolized a prized food source or frightened an infant" ("Stimulus Response" in *Psychology Today,* September 1973, p. 34).

A feature story on a female alliance among gorillas in the Yerkes Primate Research Center that appeared in the July 20, 1975, *Atlanta Journal and Constitution* was picked up by other papers around the country. Calibar and two other male gorillas were introduced to four female gorillas in the same compound so that the scientists could study their "social life." Immediately, Calibar, "who already had let his male buddies know he was the boss, set out to let the girls know he would not stand for any female sass from them," the reporter wrote in a humorous vein. Calibar "ran around the cage roughing up the girls to let them know he was stronger than they were." But the female alliance proved to be more powerful than Calibar's muscles. "Three of the girl gorillas caught Calibar in a corner . . . and beat the stuffings out of him." The report concluded with the information that

"Calibar is now in a cage by himself, trying to regain his defeated male pride and perhaps wondering where he went wrong with the ladies."

The actual status of females in a primate troop is far from the subjugated, submissive posture of "harem" captives under the rule of a despotic male. Unfortunately, even some women primatologists allow themselves to be persuaded that the myth of female inferiority is true. Such may be the case with Jane Goodall.

Goodall's data on the Gombe Stream chimpanzees does not in any way substantiate the theory of male dominance over females. Yet in her interview with Maggi Scarf, in the *New York Times Magazine,* she said she had told a student that his "feeling" about such dominance was correct; that primates do live in male-dominated groups. She then added—to the laughter of both women—that if the student was "looking for Women's Lib in the animal world" he would have to go to a "hyena society," which is "female-dominated." This jesting about a serious matter of utmost concern to women today, is not in keeping with Goodall's own data or with scientific objectivity.

Before the study of monkeys and apes became a science, fanciful tales were told by those who encountered these strange humanlike animals. John. T. Emlen, Jr., and George Schaller write,

> since du Chaillu first published his rather exaggerated accounts of gorilla life in West Africa in 1863, hundreds of articles and books have been written about them. Few of the writers were concerned with obtaining or presenting information. They were, like many modern writers, sensationalists who found self-satisfaction and financial profit in exaggeration and irresponsible prevarication. Their stories have resulted in a widespread misconception of the gorilla as a ferocious and blood-thirty beast with an amazing array of human and even superhuman traits. . . . Even scientifically-minded naturalists have been influenced by these accounts ["In the Home of the Mountain Gorilla," in *Primate Social Behavior* pp. 126-27].

Primatology has come a long way from this prescientific period. But it still has a way to go to overcome its prejudices against the female sex and its unwillingness to acknowledge the uniqueness of humans as distinct from animals.

Sociobiology and Pseudoscience (1975)

The eminent Harvard zoologist Edward O. Wilson, whose specialty is the study of insects, mainly ants, published an authoritative work on the subject in 1971 called *The Insect Societies*. That work has since been largely incorporated into a massive tome called *Sociobiology: The New Synthesis*.

With its 700 outsize pages, 120 of them devoted to glossary, bibliography, and index, the new abundantly illustrated book weighs nearly six pounds. For all that, it is retrogressive compared to his earlier work. Wilson has here amplified his entomological studies with some reports on birds, mammals, and primates in order to draw exclusively biological conclusions about human life and behavior.

Sociobiology is a word so new that it will not be found in dictionaries fifteen years old. The term implies the fusion of two sciences—sociology (or anthropology), and biology—and the correlation of relevant findings from both to shed light upon the origin and nature of human society.

But Wilson does not weave the two together. He excludes the decisive productive, social, and cultural factors that mark humans off from all forms of animal life and views all evolution, social as well as natural, as predominantly biological. This interpretation is even more narrowly reduced to "genetic evolution." Since genes make up all organic life from bacteria to human beings, Wilson perceives no qualitative jump from animal evolution to human evolution. In his view all species which aggregate into groups (and all species do) are lumped together indiscriminately as "societies." Wilson's book, despite its new prefix *socio*, is really about biology.

This attempt to explain human life in terms of animals, birds, and insects is not new. Biologism has been with us ever since Darwin set forth his theory of evolution. Once the animal origin of humans was ascertained, the Garden of Eden myth was replaced by the scientific study of the genesis of humankind. This required an examination of animal evolution and then of the socializing factors that transformed a certain branch of the higher apes into the first hominids.

Mechanical-minded thinkers, however, could not pass beyond the biological factors that led to human life. They inflated certain characteristics common to both humans and animals while underplaying or erasing the vast distinctions between them. The school of biologism gave birth to two main trends of thought: one emphasizing animal competition and the other animal cooperation to account for human competition and cooperation.

Preconditions for Humanization

The first fostered "social Darwinism," which is sometimes called the "nothing but" school. Man, its proponents said, was nothing but an animal with a few extra tricks. The catchwords "struggle for survival" and "survival of the fittest" were bandied about to buttress the thesis that animal jungle relations were carried over into the modern capitalist jungle. The proposition that "human nature never changes" meant that human nature is nothing but animal nature.

The other tendency, offended by the one-sidedness of the tooth-and-claw theorists, affirmed that not only competition but cooperation could be found in animal behavior. They pointed to the "social insects" as confirmation. This thesis was popularized by Wilson's predecessor at Harvard, W. N. Wheeler, another renowned entomologist. In 1922, after the First World War and the Russian revolution, he gave six lectures on the cooperative insects that were subsequently published in the book *Social Life Among the Insects*.

Wheeler was imbued with good intentions. He singled out the insects, he said, because "they represent Nature's most startling efforts in communal organization," and thus had "developed a cooperative communism so complete that in comparison the most radical of our bolsheviks are ultraconservative capitalists" (*Social Life Among the Insects,* pp. 5, 8). He appealed for worldwide disarmament on the basis that if such "organic cooperativeness" could exist among insects it could surely prevail among men. In place of Rousseau's "noble savage," Wheeler saw a model for superior humans in the "noble insect."

Nonetheless, Wheeler's kindly endeavor remained within the confines of biologism. In place of the notion that man is "nothing but" an animal there arose a more subtle variation of the same theme. Human society was viewed as "nothing but" an extension of insect "societies" with a few cultural features added. Soon other specialists began to jump on the bandwagon of humanizing the insects and insecticizing the humans.

Some eminent anthropologists carried forward this insect biologism. In 1953, thirty years after Wheeler's book came out, A. L. Kroeber, then dean of American anthropology, wrote in all seriousness, "Social behavior extends far back in the history of life on earth—certain insect families are much more effectively socialized than we are" (*Anthropology Today,* p. xiv). Today Wilson propagates the same theme in his *Sociobiology.*

But Wilson's net takes in far more than the "social insects." His "societies" include every species from the lowest to the highest. For example, in a caption for an illustration he uses the term "social bacterium" (*Sociobiology,* p. 392). Then came certain molds, corals, sponges, and jellyfish types. At the very top stands man. In between, along with his insect aggregates, Wilson elevates all the other species to social status. Among them are the dolphin schools, dog packs, elephant herds, lion prides, and primate troops.

In comparing the virtues of all these "societies," Wilson finds the most perfect to be not the highest but the lowest. He presents a theory of evolutionary values in reverse. It is not the vertebrates but invertebrates, such as the jellyfish, that have "come close to producing perfect societies," he says (ibid., p. 379). The ants, termites, and other social insects are "less than perfect." Least perfect are the mammals, including the primates, which are in the direct line of human ascent. "Why has the overall trend been downward?" he queries, but he cannot find an adequate answer.

However, according to Wilson, human societies "have reversed the downward trend in social evolution that prevailed over one billion years of the previous history of life." This progress is due to the fact that we have finally made our way to becoming more like the insect societies "in cooperativeness" and have even surpassed them in "communication" (ibid., pp. 379-80). This is Wilson's style in biologism.

It is not enough to add the prefix *socio* to biology to explain the distinctive attributes that have elevated our species above all others. Humans cannot be defined through biological factors alone. The same holds true for society, which is an exclusively human acquisition. Although humans retain certain features in common with the animals, once they created their own social and cultural institutions they made a drastic departure from the animal condition—and became nonanimal, or human. This has been abundantly demonstrated by scholars in such social sciences as archaeology, paleontology, anthropology, and sociology.

Even before the word *sociobiology* was coined, scholars in the humanities began correlating the relevant findings from biology and sociology, including their related sciences, to shed light on how, when, and why the great changeover occurred from animality to humanity. Among these was Gordon Childe with his classics, *Man Makes Himself* and *What Happened in History.*

These scholars began not with bacteria, jellyfish, or termites but with the primates, the immediate predecessors of humans. They explained why humanity could not have emerged from any species lower than the higher apes of the primate order. Apes, the highest species to evolve in the billion-year history of animal evolution, alone had developed the biological preconditions required for the emergence of humans.

Among these were upright posture, stereoscopic vision, the hand, the brain, and vocal organs. However, it was the freed hand with the opposable thumb that led all the rest in effecting the transformation. It was not the mandibles of insects, nor the fins of whales, nor the paws of four-footed mammals, but the hand of the highest ape species that led to tool-making and labor activities, and therewith to the transition from ape to hominid.

No other species below the level of humans can make tools in order to produce the necessities of life or to produce new needs. From the time of the first stone axe and digging stick these new needs and the means of satisfying them have advanced, at first slowly and then with astronomical speed and abundance, up to the jet plane and the spaceship.

But to accomplish this it was necessary for the first hordes of hominids to band together in social organization for collective production and mutual sharing of products and know-how. As Gordon Childe defines society, it is "a co-operative organization for producing means to satisfy its needs, for reproducing itself—and for producing new needs" (*What Happened in History,* p. 17).

Society Created by Cooperative Producers

Animals share with humans only one of these capacities: the capacity to reproduce themselves. But the production of the necessities of life and the production of new needs is exclusively human. From this standpoint we can say that production and the social organization required for its achievement mark the great dividing line between humans and animals. Thus a billion years of purely *animal* evolution was climaxed about a million years ago when the first tool-making hominids appeared on earth and set forth on their own wholly new course of *social* evolution.

Thereafter humans drastically changed the relationship between themselves and nature as contrasted with the animals. Animals are the creatures of nature, restricted to specific environments to which they must adapt themselves or perish. They are completely dependent upon what nature yields for their sustenance, unlike humans, who can cultivate the ground and produce an abundance of food. Animals are obliged to satisfy their physical needs with little or no variation in their standardized behavior patterns, while humans can roam the globe, altering their surroundings and the materials found in them for their new needs.

As social beings humans developed their minds and intellectual capacities, along with language, culture, art, and science. As they became increasingly the controllers and masters of external nature, they also changed their former animal nature into human nature.

Those who fail to see the part played by labor in the making of humankind are unable to explain how speech, language, and culture began. Wilson writes, "The great dividing line in the evolution of communication lies between man and all of the remaining ten million or so species of organisms." He even contrasts "one of the most sophisticated of all animal communication systems, the celebrated waggle dance of the honeybee," with "our own unique verbal system" (*Sociobiology,* p. 177). But he does not explain how this unique acquisition of articulate speech and language came into existence.

It grew up directly out of labor activities, out of the need for the collaborating producers to communicate with one another as well as to pass on their techniques to new generations. While our communication systems are indeed unique, the great dividing line between us and the animals originated with tool-making and labor activities.

Ants, wasps, and bees (above) form aggregates for reproductive functions. Only humans can create the social organization required for their productive and cultural activities.

Wilson carefully avoids any discussion of tool-making or production. But he is clearly on the side of those primatologists who underplay its importance in human life. They say that since primates make and use tools, just as humans do, tool-making does not provide the essential difference between them. For evidence Jane Goodall and others point to the fact that some primates have occasionally been seen to defoliate a twig and insert it under a rock to get at ants or other edibles. Wilson devotes several pages to such examples of animal "tool-using," citing nine ways in which primates can manipulate twigs, leaves, and sticks (ibid., pp. 172-75).

Primates in the wild may occasionally defoliate a twig to dig out ants, and they are capable of manipulating all sorts of objects with their hands. But this does not represent tool-making or labor activities to produce the necessities of life. Primates do not regularly depend upon twig manipulation but upon their bare hands to grasp food and convey it to their mouths. Humans by contrast are dependent upon their tool-making and productive activities for survival. If their production ceases they will perish.

Primate practices with twigs and sticks do give us a clue to the crucial importance of the freed and flexible primate hand at that critical turning point when our branch of the higher apes began to make tools and engage in systematic labor activities. But that transition from ape to human occurred only once on this planet with a special branch of the primate species a million years ago. Since then, humans have remained the only tool-makers and producers on earth, qualitatively distinct from all lower species. In other words, a billion years of purely animal evolution resulted in a revolutionary change—the departure of one species, the hominids, from its former animal conditions of existence.

Wilson's notion that human society is only a slightly improved version of insect societies by virtue of its improved means of communication is completely off base. The insect aggregate is directed to one purpose: the reproduction and perpetuation of their species. To be sure, insects have their own specialized mode of reproduction, which differs from that of the mammals. Among mammals, including the primates, the same mother that gives birth nurtures and protects the offspring. Among the insects there is a division of functions within the whole aggregate: the egg-laying female lays the eggs while other, nonbreeding, females feed and protect the grubs.

However, insects are a divergent offshoot from the main road of

The Ascent of Humans (clockwise from bottom left): from the backboned fish, to the amphibian, to the mammal, to the tree primate, to the ground primate, to humankind.

organic evolution leading to humankind—a point that Wilson fails to make clear. The point of departure for mammalian and human evolution begins with the vertebrates that grew up out of the earlier invertebrates. Originating as fish, the backboned creatures evolved into amphibians, reptiles, and then into mammals. Out of the mammals there arose the monkeys and apes, with the highest branch of apes evolving into the first hominids. Insects have no place in this line of evolution from the vertebrate fish to the backboned human beings.

The significant point is that however much the modes of reproduction differ among the various species in the animal world, there are none below the level of humans that can make tools and produce things to satisfy their needs, and generate new needs. Contrary to Wilson, insect aggregates do not furnish a prototype of human society. For if humans did no more than reproduce their kind, they would not be humans but simply primates like the chimpanzee and gorilla.

Wilson's attempt to characterize every species of animal and subanimal as "societies" is reminiscent of our primitive ancestors, who believed that animals, fish, and so forth, were organized into social clans and tribes like their own. Some thought buffalo could shoot with bows and arrows if only they had them. Others thought fish had their own territories, games, and ball parks, and that crocodiles could enter into peace negotiations with tribesmen. Primitive people were ignorant of the biological facts of life, and thus attempted to elevate animals to the human level. The Harvard professor tries to reduce humans to the animal and insect level because he disregards the social factors that make humans qualitatively different.

In Charles Darwin's day, in the struggle against the antievolutionists, it was imperative to bring out the *continuity* between ourselves and the rest of living nature. Now it is necessary to insist on the *discontinuity* between humans and animals against the biologizers, Wilson included.

In the *New York Times* advertisement of Wilson's book the headline read, "Now There's One Science for All Social Creatures." His "new synthesis" for the first time covers the "whole range of social creatures—from bacteria to termites, from monkeys to mankind," it says. This is highly misleading. A review of natural history cannot explain anything more than the preconditions for human life. It requires social history and the science of sociology to explain the origin and unique attributes of human life and culture. Wilson's "one science for all" is not a new synthesis of sociology and biology but only a new variation of an old theme—biologism—as a replacement for the scientific understanding of society.

A Crude Biologizer

Like other biologizers, Wilson used modern class and capitalist terms to describe animal and insect life. His "social insects" are divided into "castes," with a queen at the top ruling over "workers" and "soldiers." This practice is not original with Wilson; the concept of class divisions among insects has quite a long history, although Wilson only goes as far back as 1609 with Charles Butler's *The Feminine Monarchie*. What is surprising is that Wilson, whose extensive studies make him one of the world's leading authorities in advanced entomology, continues to use erroneous terms that were applied in the infancy of the science.

Almost fifty years ago Robert Briffault, citing Aristotle, Pliny, and others, showed the origin and evolution of these misleading terms. He wrote that the ancients regarded the egg-laying female as a patriarchal male and called it the "king." Corresponding to their politics, the bees were divided into "patricians" and "plebeians." When the true sexes of the insects later became known, the egg-laying female was called the "queen." By the nineteenth century the beehive was freely compared to capitalist industry—a "hive of industry" (*The Mothers,* vol. I, pp. 161-62).

In one way Wilson's biologism differs from most. He has no objection to using the term *matriarchy* and is willing to say that "matrilineal societies" exist in nature, rather than making the animal aggregates uniformly partiarchal as others do.

The term *matriarchy*, when applied to primitive society where it belongs, brings shudders to most anthropologists today. But they make no objection when the term is misapplied and the "maternal brood" in nature is erroneously called a "matriarchy." Thus Wilson feels safe in calling these maternal broods "matriarchies" since he does so entirely within his zoological framework. On this basis he finds matriarchies and matrilineal societies on every level from ants to elephants.

As Wilson points out with respect to certain insect colonies, these "castes" are all female from the "queen" down to the "workers" and "soldiers." Males are present, but only to inseminate the egg-laying female. Under these circumstances, to Wilson they are matrilineal societies (*Sociobiology,* p. 314).

Among the hoofed mammals, or ungulates, where female herds

generally keep apart from males except for the rutting season, Wilson singles out the African elephant as a striking example of a "matriarch" in charge of her daughters and granddaughters. These "female-female bonds can be assumed to last as long as 50 years," he writes. "The matriarch rallies the others and leads them from one place to another. She takes the forward position when confronting danger and the rear position during retreats." Wilson's text is accompanied by a picture of these intelligent animals in protective formation with their young. In the background are peripheral males, two fighting each other for dominance (ibid., pp. 494-97).

In another instance Wilson speaks of the "matrilineal" red deer, where a female leads the herd and another female brings up the rear. As with the African elephants, he points out, the adult females and males stay apart except during the rutting season (ibid., p. 312). Again, the caption of a picture illustrating the reconstructed "social life" of the dinosaurs of many millions of years ago reads: "A herd of females and young moves in from the left, led by an old matriarch. In the foreground two males fight for dominance" (ibid., pp. 446-47).

Among the carnivores, which stand higher in animal evolution than the ungulates, Wilson shows that the lion pride is more accurately a pride of lionesses; here too the males are peripheral.

> The core of a lion pride is a closed sisterhood of several adult females. . . . The degree of cooperation that the female members display is one of the most extreme recorded for mammal species other than man. The lionesses often stalk prey by fanning out and then rushing simultaneously from different directions. Their young, like calves of the African elephant, are maintained in something like a crèche: each lactating female . . . will permit [the cubs] of other pride members to suckle. A single cub may wander to three, four, or five nursing females in succession. . . . The adult males, in contrast, exist as partial parasites on the females [ibid., p. 504].

Above the carnivores are the primates. Wilson writes that the macaques and chimpanzees are "matrifocal" in the choice of helpers. The females band together and trust their infants to one another. He notes of the rhesus monkeys that "the mother comes to trust the females and to use them as baby sitters while she conducts foraging trips." Puzzled by this female cooperation, he asks: "Why should females care for the infants of others, and why should mothers tolerate such behavior?" (Ibid., p. 350.)

Sex Disparities

Wilson, who is so foggy about the qualitative species distinction between humans and animals, seems to be equally at sea about the sex differentiation in nature. He exclaims, "Why do the sexes differ so much?" He observes that "often the two sexes differ so much as to seem to belong to different species." Among ants and other insects "males and females are so strikingly distinct in appearance that they can be matched with certainty to species only by discovering them *in copula*." With some fishes "the males are reduced to parasitic appendages attached to the bodies of the females" (ibid., p. 318).

In certain species of insects the egg-laying female, or "queen," lives to four and a half years; but "the males, in contrast, enjoy only one to three weeks of adult existence" (ibid., p. 140). Among the ants, bees, and wasps, "males are usually discriminated against as a group. They are offered less food by the workers. . . . and in times of starvation they are frequently driven from the nest or killed" (ibid., p. 203). Among certain dance flies "the female occasionally seizes and eats the male" (ibid., p. 227). Apparently the so-called black widow spider is not unique.

It would seem that these and other facts about insects would dispel Wilson's illusion that they can be classified as "societies." Men are not mere appendages of women, rendered useless and doomed to die after insemination; they are not discriminated against or eaten by females. On the contrary, after having been trained in productive techniques by the women of the matriarchal period, they went on to become the social, cultural, and political leaders of patriarchal society. It is odd that Wilson, like Kroeber and Wheeler, should believe that insect aggregates furnish a model for human society.

The sole function of the subhuman sexes is to perpetuate the species. The male role is limited to inseminating the female. The functions of giving birth and caring for and protecting the offspring are, in the overwhelming majority of species, assumed by the females, who normally segregate themselves from the quarreling males during the maternal cycle. This uneven development of the sexes resulted in an antagonism between sex and maternity, which was carried over into the first stage of human life and had to be resolved through social means.

Wilson is well aware of this antagonism in nature. He takes

issue with Konrad Lorenz for minimizing the competitive strife in the animal world. He writes,

> The annals of lethal violence among vertebrate species are beginning to lengthen. Male Japanese and pig-tailed macaques have been seen to kill one another under seminatural and captive conditions when fighting for supremacy. . . . In central India, roaming langur males sometimes invade established troops, oust the dominant male, and kill all of the infants. . . . Young black-headed gulls . . . are attacked and sometimes killed by other gulls. . . .
>
> The evidence of murder and cannibalism in mammals and other vertebrates has now accumulated to the point that we must completely reverse the conclusion advanced by Konrad Lorenz in his book *On Aggression* [*Sociobiology,* p. 246].

According to Wilson, violence among animals of many species far exceeds that among humans in present-day society.

From evidence that has long been recognized by other scholars, Wilson concludes that "sex is an antisocial force in evolution. Bonds are formed between individuals in spite of sex and not because of it." From the broader standpoint he writes, "Social evolution is constrained and shaped by the necessities of sexual reproduction and not promoted by it." Oddly enough, he admits that this is true even of his model cooperative insect "societies." He observes that "the antagonism between sex and sociality is most strikingly displayed in the social insects" (ibid., pp. 314-15). Logical consistency is clearly not the hallmark of biologism.

The primacy of the female sex in nature and the marked differences between the sexes were long ago spelled out by Briffault in his matriarchal theory of social origins. With the exception of a few species where males are adapted to assist the females in the care of offspring, the general rule—especially among mammals—is that the females alone feed and protect their young.

Wilson himself writes that "the mother-offspring group is the universal nuclear unit of mammalian societies" (ibid., p. 456). What he fails to understand is that these maternal functions placed females in the lead at that turning point in history when the animal maternal brood was transmuted by labor activities into the human maternal clan system of social organization.

Although Wilson recognizes the primacy of the female sex in nature to a certain extent, he incorrectly refers to animal females as "matriarchs" and to their herds, packs, or prides as "matrilineal societies." The capacity of mammalian females to band together and suckle their offspring in common gives us valuable clues—along with the flexible hand and other anatomical organs—to the biological *preconditions* for human life. But to explain the matriarchy we must delve into the *conditions* required for human survival and development, and for this we must refer to other sciences, beginning with anthropology.

Male-Biased Anthropology

In anthropology, however, Wilson takes his cues from the prevailing academic schools, which are antimatriarchy and insist that the father-family and male supremacy have always existed. This obliges him to describe animal behavior in patriarchal terms that often conflict with his matriarchal terms. Thus he repeatedly refers to animal and insect reproductive broods as family "kin" composed not only of fathers and mothers but also of uncles, aunts, nieces, nephews, and cousins of all degrees. This corresponds to his view that the "nuclear family," which he regards as the "building block of nearly all human societies" (ibid., p. 553), has existed from time immemorial throughout the animal world.

In a similar vein he sees sexism as a universal trait, present in the animal as well as the human worlds. He refers to the "rampant *machismo*"—a term used today to signify male supremacy over women—that exists among some insects (ibid., p. 320). What he actually describes is not male chauvinism toward females but the fierce struggles of males against other males for dominance.

The same is true of the "violent *machismo*" he says exists in the breeding season among many other species, including sheep, deer, antelopes, grouse, lek birds, and elephant seals, where the males fight one another for dominance (ibid., p. 243). These struggles between males have no connection whatever with supremacy over females. Such supremacy did not exist in matriarchal society, much less in the animal world, where the primacy of females is so pronounced.

Wilson's notions about universal sexism converge with those of the cruder anthropologists today. In describing some male animals who keep "harems" and exercise strong "leadership," Wilson writes, "The obvious parallels to human behavior have been noted by several writers, but most explicitly and persuasively by Tiger (1969) and Tiger and Fox (1971)" (ibid., p. 287).

Elsewhere he names all five of the best-known popularizers and the most vulgar distorters of anthropology and biology: Ardrey, Morris, Lorenz, Tiger, and Fox. He commends these gentlemen for their "style and vigor." His only criticism is that "their particular handling of the problem tended to be inefficient and misleading." This comes down to an insufficient sampling of animal species to sustain their generalizations (ibid., p. 551).

Wilson is annoyed because some of the crudities of Tiger and Fox brought about a counter "feminist theory" in the book by Elaine Morgan, *The Descent of Woman* (1971). Her aquatic theory likens human females to whales and other sea mammals to show that from the beginning females have been the equals of males. Since her book also became a best seller, Wilson complains that "science" is now becoming a "wide-open game in which any number can play" (ibid., p. 29).

Indeed, once human society is biologized, anyone can play the game by inventing a fanciful hypothesis equating us with animals. Elaine Morgan's book at least has the merit of a refreshing new type of biologism compared to the stale myths of eternal male supremacy. However, despite the concessions he makes to animal matriarchs, Wilson prefers to play his overall game with the well-worn chips of the Tiger-Fox band. Thus he depicts the animal aggregates of females and offspring with the male stud attached to the group as a "harem" under the domination of a male lord and master. He calls this "polygamy" as compared to "monogamy," both of which he has projected from the patriarchal marriage institution of our times back into the animal world.

Wilson gives the game away, however, with his peculiar definitions of these marriage forms. He writes: "*Monogamy* is the condition in which one male and one female join to rear at least a single brood. It lasts for a season and sometimes, in a small minority of species, extends for a lifetime. *Polygamy* in the broad sense covers any form of multiple mating" (ibid., p. 327). Both of these definitions are misleading when applied to animal relations.

However much the marriage institution has been shaken in recent years, monogamy remains a legal term signifying property provisions for life for wife and children. The loose and casual sexual intercourse of animals is carried on without reference to legality or to any property or economic provisions. As for the polygamy of the early patriarchs, "multiple marriage" was for men only. Women had no right to divorce or any other kind of escape from their male owners.

No Monogamy in Nature

It is incorrect to use the terms monogamy and polygamy to describe the sexual practices of animals, birds, or insects. Even though some pairs of birds or a few other species may remain together longer than others, with the great majority of species there is only the act of sexual congress and no cohabitation at all between the sexes. The segregation of the sexes is far more pronounced than their fleeting unions.

Females in nature, like males, are promiscuous. This includes the higher apes. Wilson himself observes that "chimpanzee females are essentially promiscuous. They often copulate with more than one male in rapid succession, yet without provoking interference from nearby males" (ibid., p. 546). Under rigid patriarchal marriage rules, so vigorous a female would be subjected to harsh punishments.

Wilson's introduction of patriarchal and private-property relations into the animal world casts a dubious shadow over other of his interpretations of insect and animal behavior. How can one trust a zoologist who speaks of "inheritance" systems and "territorial" rights in the sense of the private ownership of real estate? For example, he writes about the black bears of Minnesota that the females "permit their female offspring to share subdivisions of the territories and bequeath their rights to these offspring when they move away or die." Male black bears "take no part in this inheritance" (ibid., p. 502).

To be sure, all animals occupy the natural habitat, or "territory," to which they are adapted and in which they find their food and mates. But there is no subdivision of real-estate properties in their world and no inheritance of property, whether from mothers to daughters or from fathers to sons.

Another example of Wilson's crude biologism and anthropomorphism is his patriarchal interpretation of the hamadryas baboons. He writes, "A female competing with a rival moves next to the overlord male, where she is in a better position to intimidate and resist attack. If she is threatened, the male is much more likely to drive her rival away than to punish her. As a result she is more likely to advance in social rank" (ibid., p. 517).

Female primates, including the hamadryas baboons, do not

have "overlords" or sultans protecting them from female "rivals" or advancing them to higher status and rank. According to Wilson, the hamadryas aggregates "contain from one to as many as ten adult females," which he calls "harems" (ibid., p. 534).

In seminatural or captive conditions a solitary female may welcome the protection afforded by a combative male who fends off other males. But so long as the females outnumber the males—as they do in the so-called harems—they have the power of numbers to protect themselves and keep the males in line. When females retreat from all males to give birth, no male follows them into their retreats.

The greatest dangers to females do not come from imaginary female "rivals" but from male assaults under captive conditions where the females have neither the power of numbers nor retreats by which to escape the males. This was demonstrated by the London Zoo experiment, when a small number of females were introduced into a large all-male colony to study their sexual habits. The results were disastrous for the females and offspring. (See *Woman's Evolution,* pp. 62-67.)

Wilson underplays the significance of this experiment. He writes, "When groups of hamadryas baboons were first introduced into a large enclosure in the London Zoo, social relationships were highly unstable and males fought viciously over possession of the females, sometimes to the death" (*Sociobiology,* p. 22).

This is an understatement. Not only were some males killed by other males, but virtually all the females and offspring were exterminated. The "unstable relations" lasted five years, during which fresh batches of females were repeatedly brought in when previous batches were killed, with no better results. In the end the experiment was declared a failure.

Wilson skims over the highly instructive lesson to be derived from the London Zoo experiment because it does not sustain his thesis about the close mating and family "kin" ties that prevail in the animal world. By pressing his erroneous patriarchal views he can declare that men resemble animals and animals men in their universal supremacy over females. He writes, "What we can conclude with some degree of confidence is that primitive men lived in small territorial groups, within which males were dominant over females" (ibid., p. 567).

Invoking the Paris professor Lévi-Strauss, he writes that "a key early step in human social evolution was the use of women in barter. As males acquired status through the control of females, they used them as objects of exchange to cement alliances and bolster kinship networks" (ibid., p. 553). What happened in the million years between the killer apes and the bartering humans Wilson does not say.

Wilson's "sociobiology" takes a step beyond the standard pattern of biologism. He has latched onto a subtler variation of the same theme—the determination of society and culture through the genes. He sees social evolution as a contest between altruism and selfishness—substitutes for what were earlier called cooperation and competition. Thus his book opens with a chapter called "The Morality of the Gene." He locates altruism in the genes and believes it is through genetic selection—not social progress—that altruism will triumph in human relations.

In fact, altruism and selfishness are exclusively characteristics of human relations and the moral judgments made of human behavior in the course of history. Animal competition or "selfishness" was conquered in the primitive human world through the institution by the women of a matriarchal, communal society. Altruism came into existence as men learned to live and work together as brothers, interchanging the necessities and comforts of life.

With the downfall of the primitive commune and its replacement by patriarchal class society, that aboriginal altruism was subverted and a new kind of selfishness came into existence out of the greed bred by the lust for private property. But this subverted human nature will once again be changed when capitalism is replaced by socialism.

Wilson's "Genetic Capitalism"

Wilson has a different view. He thinks it possible that the social classes in capitalist society are formed through genetic differentiation. He writes, "A key question of human biology is whether there exists a genetic predisposition to enter certain classes and to play certain roles. Circumstances can be easily conceived in which such genetic dfferentiation may occur" (ibid., p. 554). Does he mean that a capitalist is genetically preordained to play that role while a sanitation worker is genetically programmed for that work? And does he mean that the female sex is genetically programmed to be inferior?

Wilson creeps around the question. He is attracted to Dahlberg

(1947), who "showed that if a single gene appears that is responsible for success and an upward shift in status, it can be rapidly concentrated in the uppermost socio-economic classes." Short of gene manipulation, this can only be done through controlled mating; capitalists intermarrying with the rich, sanitation workers with the poor.

The sinister implications in this school of genetic determinists have already been denounced. From Hitler on to the present there have been advocates of controlled mating to keep out the non-Aryan and other "foreign" genes and maintain a pure and high-class stock for the master race and the master class.

Wilson shies away from such a conclusion. He writes, "Despite the plausibility of the general argument [of Dahlberg], there is little evidence of any hereditary solidification of status." In his view there are many other "pathways of upward mobility." For example, "The daughters of lower classes tend to marry upward." Presumably the inequalities of capitalism will be gradually eliminated through this constant reproduction of high-class genes in the children of the poor daughters who marry rich husbands.

Pollyanna Wilson writes, "over a period of decades or at most centuries ghettos are replaced, races and subject people are liberated, the conquerors are conquered" (ibid., p. 555). According to Wilson's doctrine of the *"genetic evolution of ethics"* (ibid., p. 563, his emphasis), there is no need for a social revolution to get rid of an oppressive capitalist system. Presumably the problem will be taken care of when all the poor daughters have married all the rich husbands. Then everybody will have the same high-class genes. This has been aptly characterized as "genetic capitalism."

Wilson's genetic interpretation of human evolution runs directly counter to those sociologists and others who say that human behavior is learned behavior, that it is not derived from the genes but from their own productive, social, and cultural activities. He criticizes Dobzhansky (1963), who stated that "culture is not inherited through genes, it is acquired by learning from other human beings. . . . In a sense, human genes have surrendered their primacy in human evolution to an entirely new, nonbiological or superorganic agent, culture." Wilson thinks "the very opposite could be true." He calls for a new "discipline of anthropological genetics" to prove it (ibid., p. 550).

Wilson thus diverges from two molecular biologists, Alan Wilson and Mary Claire King, who challenged the notion that genetic mutations are responsible for the elevation of humans above the animals. The *New York Times* of April 18, 1975, reported that "the scientists suggest that some other form of mutation must be operating to have produced the obviously vast differences between people and chimpanzees."

Indeed, that "mutation" was nothing other than the qualitative jump that occurred when the first humans acquired their new mode of survival and development through tool-making and labor activities. This is why society is an exclusively human phenomenon.

The two sciences of biology and anthropology are of the greatest importance in understanding the origin, evolution, and meaning of social life. But Wilson, instead of advancing these sciences, is turning them backward. In the name of promoting a new branch of science he has descended to the level of the Ardrey-Morris-Lorenz-Tiger-Fox obscurantists. He is lending himself to indoctrinating unwary and trustful readers with reactionary ideas and encouraging pseudoscientists to jump on his bandwagon.

As Alexander Cockburn, staff writer of the *Village Voice,* warned in his July 28, 1975, review of *Sociobiology:* "Now we have 'sociobiology' and the probability of a new terrible wave of zoomorphist rubbish. Brace yourselves."

An Answer to 'The Naked Ape' and Other Books on Aggression (1970)

Since the early 1960s the United States, the most powerfully armed nation on earth, has been conducting an onslaught against Vietnam, a tiny nation far from its shores. This long drawn-out, genocidal war has produced wave upon wave of revulsion among the American people.

Massive, unprecedented antiwar demonstrations have been accompanied by an intense interest in the root causes of military conflict. Many Americans who once believed that wars were waged only to "safeguard democracy" rightly suspect that they have been hoodwinked. They are coming to see that the only gainers from such conflicts are the monopolists, who seek to safeguard their empire and expand their power, profits, and privileges through them. Thus a political awakening is taking place with regard to the real causes of imperialist aggression, which are embedded in the drives and decline of the capitalist system.

In the same time period a set of writers has come to the fore whose books present a wholly different view of the causes of organized warfare. They claim that man's biological heritage and his "killer" instincts are responsible for wars, absolving the predatory capitalist system of all responsibility. Their paperbacks are bought by the hundreds of thousands and have been high on the best-seller lists. They obviously influence the thinking of many readers who are anxiously searching for answers to the problems of war and other social evils.

The principal figures among these capitalist apologists have produced six such books in the decade. The pacesetter is Robert Ardrey, who brought out *African Genesis* in 1961 and its sequel, *The Territorial Imperative,* five years later. A third, *The Social*

Contract, has just been published. Konrad Lorenz published *On Aggression* in 1963, which was translated into English in 1966. In 1967 *The Naked Ape*, by Desmond Morris, appeared, followed two years later by its companion, *The Human Zoo.*

The authors come from different countries and have dissimilar backgrounds. Ardrey was an unsuccessful playwright who became a dabbler in anthropology. Lorenz is an Austrian naturalist, sometimes called the "father of ethology"—the science of animal behavior in the wild—who specializes in the study of the greylag goose and certain other bird and fish species. The Englishman Morris was formerly curator of mammals in the London Zoo.

However much these writers differ in background, training, and temperament, they agree that modern wars are not brought about for economic and social reasons but stem from the biological aggressiveness of human nature.

Their method consists in obliterating the essential distinctions that separate humans from animals and identifying the behavior of both through gross exaggerations and misrepresentations of the part played by instincts in human life. They argue that since mankind came out of the animal world, people are at bottom no better than animals; they are inescapably creatures of their biological impulses. Thus modern warfare is explained by man's "innate" aggression.

This extension of animal aggressiveness to account for imperialism and its military interventions is absurd on the face of it. No animal has ever manufactured an atom bomb, and there are no apes standing ready to hurl them and blow up the planet. The small group of aggressive men who control the nuclear warheads are not in the zoos or the forests but in the White House and Pentagon.

To equate animal behavior with imperialist warmaking is to slander not only animals but the vast majority of humans who wish only to live in peace. The Vietnamese have not threatened or invaded the territory of the United States; the opposite is the case. And the average GI has so little warlike animosity for these distant "enemies" that it requires heavy pressure and unremitting patriotic indoctrination to convince him that he must become aggressive against them.

To the new school of writers, however, wars are not made by big business and its agents in Washington; the real culprit is the ape nature of man. With this biological fig leaf, they attempt to

No ape ever hurled an atom bomb.

cover up the criminal course of the imperialists, and dump responsibility for their aggressions upon "man" in general.

These writers refuse to recognize that, while humankind has grown out of the animal world, we are a unique species which has outgrown animality. A whole series of distinctive attributes divides us from all lower species. Humans alone have the capacity to *produce* the necessities and comforts of life; humans alone possess speech and culture; humans, therefore, make their own history. The laws of social evolution, applicable to humanity alone, are fundamentally distinct from the laws of organic evolution applicable in nature.

This point is made by the eminent paleontologist, George Gaylord Simpson, in his book *The Meaning of Evolution*:

> The establishment of the fact that man is a primate, with all its evolutionary implications, early gave rise to fallacies for which there is no longer any excuse (and never was much). . . . These fallacies arise from what Julian Huxley calls the "nothing but" school. It was felt or said that because man is an animal, a primate, and so on, he is *nothing but* an animal, or *nothing but* an ape with a few extra tricks. It is a fact that man is an animal, but it is not a fact that he is nothing but an animal. . . . Such statements are not only untrue but also vicious for they deliberately lead astray enquiry as to what man really is and so distort our whole comprehension of ourselves and our proper values.
>
> To say that man is nothing but an animal is to deny, by implication, that he has *essential* attributes other than those of all animals. . . . His unique nature lies precisely in those characteristics that are not shared with any other animal. His place in nature and its supreme significance to man are not defined by his animality but by his humanity [pp. 282-83].

According to Simpson, man represents "an absolute difference in kind and not only a relative difference in degree" from all animals. Ardrey, Lorenz, and Morris are clearly at odds with these statements on the qualitative distinctions between humans and animals.

The crudest of the three is Robert Ardrey, who reduces science to fiction writing. An adroit name-dropper, he sprinkles his books with references to prestigious scientists, to endow his work with their sanction. He does this, for example, with Simpson, who is far from sharing Ardrey's views about mankind.

Man is only a "fraction of the animal world," says Ardrey, and human history no more than an "afterthought" of natural history. We are not, therefore, "so unique as we should like to believe" (*African Genesis*, p. 9). This is exactly the opposite of the views expressed by Simpson on the subject.

Killers and Capitalists

Ardrey's books are designed to demonstrate not only that man is a born killer, a "legacy" bequeathed by our killer-ape ancestors, but that animal nature is also at the bottom of the lust for private property. He takes exception to Darwin's observation that male animals compete and fight for sexual access to females in the mating season. According to Ardrey, animals, like people, compete and fight for the private ownership of property which begins with one's own territory. This is the central theme behind his "territorial imperative."

To substantiate his thesis he cites a bird specialist who "observed throughout a lifetime of bird watching, that male birds quarrel seldom over females; what they quarrel over is real-estate." The females, for their part, are sexually attracted only to males possessing property. "In most but not all territorial species," we are told, "the female is sexually unresponsive to an unpropertied male" (*The Territorial Imperative*, p. 3). A mocking-bird, it seems, can get a mate only after having fought for and won sufficient holdings in property.

Highlighting this absurdity, Ardrey further assures us in *The Territorial Imperative* that "many animals," such as lions, eagles, and wolves, "form land-owning groups." He makes no distinction between the use of land, sea, or air by creatures in nature for their habitats, and the exclusive private ownership of land and other resources by rent-collectors. Thus he concludes, "Ownership of land is scarcely a human invention, as our territorial propensity is something less than a human distinction."

According to Ardrey, man has inherited his greedy proclivities from his ape ancestors, and this legacy explains human "killer" instincts in defense of possessions and territory. This justifies not only the capitalist way of life but also the imperialist aggressions that are waged by the U.S. to maintain its system. Ardrey thereupon appeals for a less negative attitude on the part of Americans today toward war, urging them not to be swayed by those who despise wars and war-makers.

"Generals in the time of my growing up were something to be hidden under history's bed, along with the chamber pots," he complains. "Anyone who chose the army for a career was a fool or a failure." Indeed, after the First World War, "certain words almost vanished from the American vocabulary, among them such fine patriotic words as *'honor'* and *'glory.'*" And he sorrowfully adds, "Patriotism, naturally, was the last refuge of the scoundrel."

Bent on changing this attitude, Ardrey warns that the same "territorial imperative" that is embedded in our instincts likewise motivates the "enemy." So if we are to save ourselves and our property we must fight, fight, fight. He writes:

> The territorial imperative is as blind as a cave fish, as consuming as a furnace, and it commands beyond logic, opposes all reason, suborns all moralities, strives for no goal more sublime than survival. . . . But today's American must also bear in mind that the territorial principle motivates all of the human species. It is not something that Americans thought up, like the skyscraper or the Chevrolet. Whether we approve or we disapprove, whether we like it or we do not, it is a power as much an ally of our enemies as it is of ourselves and our friends [ibid., p. 236].

What are we to say to this most unnatural history? It is obvious that living creatures congregate in specific habitats on the land or in the sea which provide them with food and mating grounds. But these habitats are not "territories" in the sense of landed estates that they permanently own. It is also true that animals may become aggressive in the struggle to satisfy their basic needs. But they are just as capable of tolerating one another's presence in a common habitat as they are of squabbling over any given spot at any particular time.

Aggressiveness in defense of a habitat is imposed upon animals, because for survival each species is adapted to the particular food and climate of specific areas. Thus, trooping animals may defend the region occupied by the group; solitary animals defend only the particular spot each occupies at any given time. In all cases, the "imperative" is not for "territory" but for satisfying the most basic needs of the animal within the restricted framework of its particular living space.

Conditions of life are entirely different in the human world, however, where humankind is not chained to any special food or climate and can produce what is needed anywhere on the globe. Unlike the polar bear, which cannot live in tropical Africa, or the tropical ape, which cannot survive in icy Newfoundland, human beings can roam and inhabit the whole planet, together producing and sharing the necessities of life. Humans can act consciously and collectively to eradicate war once they become aware of its causes.

More to the point, the capitalists are not so much interested in protecting "their" territory, as such, from alleged enemies; what they want to maintain at all costs is "their" system of exploitation. That is why the United States, while fighting against the "enemy" in Southeast Asia, also has military bases on other peoples' territories all around the globe. A capitalist ruling class will even temporarily yield sovereignty over its territory, if need be, as the French monied men did to Hitler during the Second World War, to preserve their properties from the insurgent masses.

The American people do not decide who their enemies are; these are singled out for them by the shifting needs of the capitalists. During the Second World War the Germans and Japanese were the enemies whereas the Soviet and Chinese allies were friends. Since then these respective nations have been switched as friend and foe. What has changed is not the territorial relations but the diplomatic and strategic aims of American imperialism. Its propaganda machine tells the country who is to be hated and who is to be liked at any given time. Contrary to Ardrey, there is nothing instinctual in these attitudes; all of it is learned behavior, instilled by the ruling class.

Lorenz and Morris, who, unlike Ardrey, have some claim to the title of scientists, go as completely wrong when they try to biologize history. This is as great an error as it would be to reduce biology and botany to chemistry and physics, even though animal and vegetable life have a physicochemical origin and basis. In the case of human life it produces grotesque distortions of the truth.

Desmond Morris is particularly crude in this respect. "I am a zoologist and the naked ape is an animal. He is therefore fair game for my pen," he declares in his first book, *The Naked Ape*. To this zoo keeper, man differs from the ape by virtue of two amplified biological organs, a bigger penis and a bigger brain, and because our species is "naked" while apes are hairy. Nothing essential has been altered by humans either in themselves or their society; they were and still remain the creatures of their ape instincts: "So there he stands, our vertical, hunting, weapon-toting, territorial, neotenous, brainy, Naked Ape, a primate by ancestry, and a carnivore by adoption, ready to conquer the world . . . for all his environment-moulding achievements, he is still at heart a very naked ape" (p. 48).

These writers, who see no qualitative distinction between man and ape, ignore the extent to which man himself has changed in the course of his million-year history. People today who are only now becoming aware of the social jungle that has been foisted upon them by the capitalists are not the same as the people of precivilized society who conquered their animal heritage and conditions of life, reconstituting themselves into the tribal brotherhood of men. Indeed, the very existence of that primitive system of collectivism and their cooperative relations testifies to how profoundly men were emancipated from their earlier brute instincts.

Instincts or Learned Behavior

The propostition upon which Ardrey, Lorenz, and Morris build their case for the innate aggressiveness of mankind, i.e., that humans are governed by irrepressible, unmodified, inherited instincts, is today rejected by most authoritative scientists. Let us examine this aspect of the matter.

The degree to which humans have shed their original instincts is so great that most of them have already vanished. A child, for example, must today be taught the dangers of fire, whereas animals flee from fire by instinct. According to anthropologist Ralph Linton, instincts, or "unlearned reactions," have been reduced to "such things as the digestive processes, adaptation of the eye to light intensity and similar involuntary responses." He adds: "The fewer instincts a species possesses, the greater the range of behaviors it can develop, and this fact, coupled with the enormous capacity for learning which characterizes humans, has resulted in a richness and variety of learned behavior which is completely without parallel in other species" (*The Tree of Culture,* p. 8).

Except for reactions in infants to sudden withdrawals of support and sudden loud noises, Ashley Montagu likewise denies that any significant aspect of human behavior is purely instinctive; all of it is conditioned by life experiences. Furthermore, as animal experiments and domestication disclose, many of the reactions of living creatures below the level of humankind, which have been conventionally classified as instinctive, can be considerably modified by social and environmental conditioning.

Lorenz, who is more prudent and scholarly than Morris, is embarrassed by his colleague's crudity. Although he upholds the thesis that man is subject to his animal instincts, he acknowledges that people are set apart from the animals by their possession of culture and language. "That's why," he commented in an interview in the July 5, 1970, *New York Times Magazine,* "I don't like my friend Desmond Morris' title, 'The Naked Ape.'" Morris, he says, disregards the fact that man is "an ape with a cumulative tradition." But the mere existence and weight of such a tradition in social development proves that mankind is human, not ape!

Unable to grasp the full import of this fact, Lorenz sides with Morris in the matter of the innate aggressiveness of humans. To him there is no essential difference between a cockfight and a nuclear war; the one follows in a continuous evolutionary sequence from the other. There is, he says, "the alarming progression of aggressive actions ranging from cocks fighting in the barnyard to dogs biting each other, boys thrashing each other, young men throwing beer mugs at each other's heads, and so on to barroom brawls about politics, and finally to wars and atom bombs" (*On Aggression,* p. 29).

Note how Lorenz leaps from animal fights to human quarrels, disregarding the decisive differences between them. Then, on the human level, he refuses to distinguish between the petty personal encounters of people and the massive military operations conducted by governments, in which men are ordered to kill in cold blood other men they have never even seen before, much less had any personal quarrel with.

Animal fights, personal squabbles, and imperialist wars are all dumped into the same sack to substantiate the falsification that humans are nothing but animals and have never passed beyond that stage of development. This theme is only a variation of the tiresome old argument that "you can't change human nature"—another piece of capitalist propaganda designed to avert revolutionary change in our social system. Their special twist is that "you can't change animal nature" since in their view humans are nothing but animals. History, however, demonstrates that just as the ape became man, so did man radically transform his ape nature and convert it into human nature.

Furthermore even this human nature has changed drastically in the course of social history, and will continue to acquire new and different traits as humanity begins to emancipate itself from capitalist thralldom. What man needs to throw off today is not animal nature, which he shed a million years ago; rather, he must throw off capitalist nature, which has been imprinted into man's conduct and psychology by this society.

This is precisely the point that the "instinctual" school of theoreticians seeks to gloss over or cover up. They fear that an acknowledgment of a changing human nature logically clears the way for a radical change in our social system. Thus Lorenz, who is most forthright in this respect, is careful to dissociate himself from the position of Marx and Engels.

In the *Times* interview he said, "Marx was very aware of the need to conserve the whole heritage of culture. Everything he said in *Capital* is right, but he always made the error of forgetting the instincts. For Marx the territorial instinct was only a cultural phenomenon."

But the founders of scientific socialism were completely right in rejecting the "instinctual" approach to social history. As they pointed out, the main motor forces in human progress are not biological but social. Humans possess that crucial characteristic which no other species possesses: the capacity to labor and develop the forces of production. Laboring humanity has the ability to anticipate, imagine, reason, pursue goals, and advance the whole sphere of culture. All this not only gives humans increasing control over their own lives and destinies, but also constantly modifies their own human nature. The renowned archaeologist Gordon Childe wrote on this point:

> In human history, clothing, tools, weapons, and traditions take the place of fur, claws, tusks, and instincts in the quest for food and shelter. Customs and prohibitions, embodying centuries of accumulated experience and handed on by social tradition, take the place of inherited instincts in facilitating the survival of our species. . . . it is essential not to lose sight of the significant distinctions between historical progress and organic evolution, between human culture and the animal's bodily equipment, between the social heritage and the biological inheritance [*Man Makes Himself,* p. 20].

The irreconcilable differences between the two schools of thought on the nature of aggression in history have more than an academic or literary interest. To say that man is governed by his ape nature and that humans are born mass murderers has important political consequences. It diverts attention from the real instigators of war, the capitalist magnates, and leads people to blame themselves for their "evil" instincts. This self-blame feeds a despairing, apathetic attitude and produces a fatalistic outlook. It tends to dissipate the social anger of masses of people who can band together in revolutionary action against those who are really to blame—the dangerous war-makers.

This mood is explicit in both Morris and Lorenz, who, seeing no revolutionary solution to capitalist-made problems, present prophecies of doom. Morris believes "there is a strong chance that we shall have exterminated ourselves by the end of the century." Lorenz is equally pessimistic and says in *On Aggression* "an unprejudiced observer from another planet, looking upon man as he is today, in his hand the atom bomb, the product of his intelligence, in his heart the aggression drive inherited from his anthropoid ancestors, which this same intelligence cannot control, would not prophesy long life for the species" (p. 49).

Marxists do not deny that all humanity is threatened with extermination by the nuclear arsenal and other death-devices controlled by the overkillers in Washington. But we believe that working men and women and their allies can be aroused and organized to take economic, military, and political power away from the capitalist atom-maniacs and thereby eradicate the causes of war. This conviction that a socialist revolution can and will put a permanent end to imperialist slaughters is the basis for Marxist optimism—as against the prophets of doom of the "instinctual" school.

The Critics Speak Out

There has been no lack of competent critics to challenge Ardrey, Lorenz, and Morris for drawing sweeping and reactionary conclusions about humans on the basis of limited, specialized, specious, and erroneous data about animal life. These scholars reject the premise that humankind is the blind creature of instincts. Most of them agree that instincts have long been supplanted by learned behavior as the dominant factor in social and cultural life. For those who may be unaware of the broad scope of the criticisms, here is a brief summary of the views of many well-known naturalists, anthropologists and sociologists who have taken issue with these writers.

The pacesetter was Marshall Sahlins, University of Michigan anthropologist, who reviewed *African Genesis* in the July 1962 *Scientific American.* "Ignoring the million years in historical development of cultural forms," he wrote, "Ardrey typically takes as *human* the conditions he finds at hand, reads them into vertebrate sociology and so accounts biologically for human behavior."

Indeed, Ardrey makes a double error in methodology without knowing that he is doing so. First he takes the behavior of human beings in capitalist society as *natural* and falsely applies it to animal behavior. Then he illegitimately projects this invalid interpretation of animal behavior back on to "man" in general. This enables him to obliterate the crucial distinctions between the natural animal and social mankind.

Following Sahlins, many other criticisms were published in the *New York Times Magazine*, scientific journals, and other media, bearing down heavily on the falsification that wars are virtually implanted in man's genes. In 1968 Ashley Montagu compiled fifteen articles specifically directed against Lorenz and Ardrey, in the anthology *Man and Aggression.*

These critics conduct their polemics along two lines. First, they assail and expose the dubious and misleading data offered in the name of science by Lorenz and Ardrey as more fictional than factual. Second, they are incensed by the thesis that wars are unavoidable because of the innate depravity of man as an instinctual killer. They point out that animals that kill for food act simply to satisfy their hunger; they are not war-makers. Nor were primitive peoples war-makers.

"Organized warfare between states is, of course, a very modern human invention," says the British anthropologist Geoffrey Gorer. The raids and skirmishes of precivilized peoples cannot be compared either in quantity or quality with the massive wars between nation-states in our times. Gorer summarizes Ardrey's "oversimplifications, questionable statements, omissions and plain inaccuracies" in stinging terms:

> Ardrey shows only the most superficial knowledge of contemporary events, practically no knowledge of the history of the old world or of contemporary sociology and social anthropology. His categories and preferences are bound to give comfort and provide ammuniton for the radical Right, for the Birchites and the Empire Loyalists, and their analogues elsewhere. . . . *The Territorial Imperative* demands a wrapper: "Handle carefully; Read with critical scepticism" ["Ardrey on Human Nature: Animals, Nations, Imperatives," in *Man and Aggression*, p. 82].

Some of the critics are gentler with Lorenz, who has made certain contributions to natural science. But they do not exculpate him for resorting to pseudoscientific arguments to buttress the myth that war-making is innate. Further, they question his qualifications as an authority on either primate or human behavior.

Lorenz is not a student of anthropoids that stand in the direct line of human ascent, nor even of the mammalian species. He has studied only birds and some fish—creatures which are far removed from mankind in the sequence of evolution. J. P. Scott, of Bowling Green (Ohio) University, says that Lorenz knows little outside his limited field; that he is "a very narrow specialist who primarily knows the behavior of birds and particularly that of ducks and geese on which his book has an excellent chapter" ("That Old-Time Aggression," in *Man and Aggression*, p. 52).

Similar criticisms were made at an international meeting held in Paris in May 1970 under the auspices of UNESCO (The United Nations Educational, Scientific, and Cultural Organization) where a score of scientists discussed the problem of aggression and war for a week. According to a report in the May 23, 1970, *New York Times*, they unanimously opposed the views of Lorenz and Morris that aggression is innate, inevitable, and even beneficial. They state that aggressive behavior is learned. People act violently because they have been taught to do so or are made to do so, not because they are born or ordained to be aggressive toward their fellow men.

Dr. Adeoye Lambo, director of the Behavior Science Research Institute at Ibadan, Nigeria, gave several examples of societies where aggressiveness in young children is consistently rewarded, to illustrate how aggressiveness is learned rather than instinctive. Several other participants pointed out that a murder or some other act of violence takes place on American television screens every eight seconds. Television also shows daily newsreels of the violence committed by the colossal United States military machine in Southeast Asia.

Professor Robert A. Hinde, director of Animal Behavior Studies at the University of Cambridge, said that both Lorenz and Morris are "very ignorant of the major chunk of literature about both animals and man." He said Lorenz reads nothing outside his specialty, and "his emphasis upon the inevitability of aggression is a gross exaggeration." He branded Morris's two books a "dangerous intertwining of fact and fiction."

These scholars and scientists are especially concerned about the damaging effect such ignorant and irresponsible assertions can have upon the millions of people who accept them as scientific gospel. As Sally Carrighar, the British naturalist, says, a social evil can only be eradicated if its true causes are recognized. But "the incentive to do it is lacking while people believe that aggression is innate and instinctive with us" ("War Is Not in Our Genes," in *Man and Aggression,* p. 50). And the economist Kenneth E. Boulding correctly stresses that "human aggression and human territoriality are products of social systems, not of biological systems. They must be treated as such" ("Am I a Man or a Mouse—or Both?" in *Man and Aggression,* p. 88).

A number of these critics recall that there is nothing new in this "tooth and claw" approach to human history. The ideas propounded by Ardrey, Lorenz, and Morris are a re-edition in modern dress of the social Darwinism that was propagated in conservative circles in the last part of the nineteenth century and up to the end of the First World War, when it faded away.

Ralph Holloway notes that the phrase "social Darwinism" never appears in *The Territorial Imperative.* "Too bad," he remarks, "for that is essentially the message of the book. Ardrey is uninformed if he thinks that there have never been attempts to reduce human group behavior to a few animal instincts" ("Territory and Aggression in Man: A Look at Ardrey's Territorial Imperative," in *Man and Aggression,* pp. 97-98).

Neo-Social Darwinism

Every epoch-making discovery can be perverted by the masters of class society and their spokesmen-servants. The capitalists, for instance, misuse machinery which is designed to lighten man's work by making humans into sweating appendages of the machine. Darwin's findings on the origin of species and the theory of evolution, which revolutionized the study of biology and threw light on the genesis of humankind, have been similarly perverted. Conservative ideologues misapplied them to the nineteenth-century social scene as a rationale for capitalist competitiveness, greed, and inequality.

The catchwords of "struggle for existence," "natural selection," and "survival of the fittest" were invoked to uphold the practices of *laissez faire*—let things run their course as they are, and the fittest will survive. This gave the sanctity of natural law to the social jungle created by capitalism at home and to its wars and territorial conquests in foreign lands.

T. K. Penniman, the British historian of anthropology, described this gospel as follows:

> Imperial developments appeared to show that the "lesser breeds without the law" were bound to go to the wall, and that such events were but the working of the law of nature. . . . The idea that one nation subdues another or annexes territory because it is superior, or that a man who gains more ease and money for less work than another, is therefore the fitter to survive and progress, are ideas begotten, not of Darwin, but of the competition for mechanical efficiency . . . people reduced to fighting for a living wage, or those who contemplated the struggle, must give the palm not to those who could take pride in what they made, or did, but to those who most successfully exploited their fellows [*A Hundred Years of Anthropology*, pp. 94-95].

The new social Darwinists have refurbished these discredited doctrines to again eternalize bourgeois relationships and justify imperalist violence. Ashley Montagu says, "There is nothing new in all this. We have heard it before. . . . As General von Bernhardi put it in 1912, 'War is a biological necessity. . .' " ("The New Litany of 'Innate Depravity,' " in *Man and Aggression,* p. 10).

One example from Lorenz should suffice to show how they revive social Darwinism. He equates the intraspecies competition among animals for food and mates with the socioeconomic competition of men today. Competition is indeed the hallmark of capitalism. The big aggregations of capital push the weaker to the wall, and workers are forced to bid against one another for the available jobs. But Lorenz views this capitalist-made competition as the result of inborn animal attributes.

"All social animals are 'status seekers,' " he informs us in *On Aggression.* Birds, like men, compete with one another for status and possessions, and the "strongest" or "fittest" wins out over the "weakest or least fit." Thus there are "high-ranking" jackdaws who have more status and wield "authority" over the lowly jackdaws who lack both status and authority.

To Lorenz there is great "survival value" in this "pecking order" of man, bird, and beast, providing the weaker submit to the stronger. "Under this rule every individual in the society knows which one is stronger and which weaker than itself, so that everyone can retreat from the stronger and expect submission from the weaker, if they should get in each other's way." Every boss today would certainly like to establish this rule with regard to the workers. Unfortunately for him, they are not birds or beasts—but men and women who can organize and fight back.

It is true that a wasteful method of species survival and development prevails in nature where, under conditions of limited food and space, competition prevails and the less fit are eliminated to the benefit of the fittest.

But such wasteful methods are unnecessary in human society today, where people can plan their lives and control their own destinies—once they get rid of the exploitation and anarchy of capitalism. As Engels commented, "Darwin did not know what a bitter satire he wrote on mankind, and especially on his countrymen, when he showed that free competition, the struggle for existence, which the economists celebrate as the highest historical achievement, is the normal state of the *animal kingdom*" (*Dialectics of Nature,* p. 19).

Racist and Sexist

Prejudices of a feather flock together. So it should come as no surprise that those who degrade humanity to the animal level are also racist and sexist in their outlook. Whereas Ardrey denies that male birds or animals fight over anything as unimportant as females but rather fight over real estate, Lorenz takes a different tack. He says that females are "no less aggressive than the males," and in particular display hostility toward members of their own sex—presumably just as women do in competitive capitalist society.

This generalization is based on observations of certain rare fish, such as the East Indian yellow cichlids and Brazilian mother-of-pearl fish, where not only are males hostile to males, but females are apparently unfriendly to females.

It is well known that in many species, above all the mammals, females will fight in defense of their offspring. Males, on the other hand, fight one another for sexual access to females. This trait is not duplicated in the female sex. A female fighting another female for access to males is conspicuous by its absence in the animal world. In herding species, one bull is quite sufficient for a herd of females, and a "pride of lions" is composed of a pack of lionesses to which usually only a solitary adult male is attached. Lorenz does not make clear the considerable differences involved in these types of aggression on the part of the animal sexes.

What is worse, he uses certain exceptional phenomena in nature as the basis for drawing sweeping conclusions about women in our society. Because certain female cichlids eat the male at their "marriage feast" and some show unfriendliness to other females, Lorenz draws from this a pattern of human behavior. He offers the following illustration:

While there was still a Hapsburg monarchy and well-to-do women had servants, his widowed aunt never kept a maid longer than ten months. To be sure, his aunt did not attack or eat the maids; she merely fired one and hired another. Her conduct, however, presumably testifies to the everlasting, innate aggressiveness of females toward other females. Lorenz mistakes the class-conditioned temper and capriciousness of a woman with her servants in capitalist society for female aggression against other females in nature, which is exceedingly rare.

Desmond Morris displays a much more profound animosity toward women than does the paternalistic Lorenz. He informs us that the beauty aids purchased by women are only modern adaptations of the "sexual signalling" of our animal ancestresses. By implication all females, both animal and human, are unattractive and ugly to males and therefore must resort to sexual lures.

As a zoo-man, Morris must known that while humans can mate all year round, animal mating is restricted to the estrus or sexual seasons. Both males and females are quiescent in the nonestrus seasons. It is only when the next sexual season opens that males are again reactivated sexually, and this occurs in response to the "sexual signalling" of the females. For it is the females who determine the opening of the sexual season. But Morris equates this natural phenomenon with the multibillion-dollar cosmetic and fashion industries in capitalist society by which the human female is assisted in the competitive struggle to snag her man.

He spells this out in considerable detail. From the "padded brassiere" to improve "sagging breasts" and the "bottom-falsies" for "skinny females" to the lipstick, rouge, and perfume—these and other devices enable women to entice the men they are after. And he pumps sex into his sexist book by devoting many pages to spicy accounts of the private parts and private lives of primate females and the kinds of erotic stimuli that move naked apes into their body-to-body contacts, with added tidbits on primate voyeurism and prostitution.

Ardrey, the outspoken jingo, is likewise the least disguised racist and sexist. In the title of his book, *African Genesis,* he popularizes the fact established by scientists that mankind had its origin a million years ago not in Asia, as previously thought, but in Africa. This highly significant fact could be used to help shatter the myth of African inferiority which has been peddled by white supremacists. If mankind had a single point of origin in Africa, it follows that, regardless of race or nation, we are all ultimately descendants of the Africans, who were the creators of the first social organization of humankind.

But this is not Ardrey's interpretation of our African genesis. According to him, it is precisely this heritage which taints us with the "killer-ape" instincts from which "man" has never recovered. This is the same old racist slander in a somewhat different form. It is reinforced when he refers to "a troop of brown lemurs in a Madagascar forest" in the same context as the Japanese attack on Pearl Harbor, implying that non-Caucasians are not quite human.

Ardrey loves white South Africa, which, despite a "degree of tyranny," has attained "peaks of affluence, order, security and internal solidarity rivalled by few long-established nations." By contrast, he states in *The Territorial Imperative,* the Black African states "stagger along on one side or the other of the narrow line between order and chaos, solvency and bankruptcy, peace and blood."

Ardrey's male supremacism matches his white supremacism. He wants women to stay in their place, which, for him, is the middle-class white home and family. He cannot understand why these favored housewives are discontented. Why, he complains querulously, does she have a "rowdy approach to the boudoir which will bring her nothing but ruin"; why does she "downgrade the care of children as insufficient focus for feminine activity," and why does she desire "masculine expression" in social life for which she possesses no "instinctual equipment?"

His answer to these questions is most revealing:

> According to every American ideal . . . she lives in a feminine Utopia. She is educated. She has been freed of the dust-mop cage. No social privilege is denied her. She has the vote, the bank account, and her entire family's destiny gripped in her beautifully manicured

> hands. Yet she is the unhappiest female that the primate world has ever seen, and the most treasured objective in her heart of hearts is the psychological castration of husbands and sons [*African Genesis*, p. 165].

There it is, spread out for everyone to see. Man is a killer-ape, and woman is a sneaky, nasty primate that castrates men.

These neo-social Darwinians are pushing the most pernicious prejudices of class society under the label of biological and anthropological "science." The enterprise is highly lucrative for them and their publishers. But unwary readers should be warned that they are receiving large doses of poison in the same package with only a few facts.

Lionel Tiger's 'Men in Groups': Self-Portrait of a Woman-Hater (1977)

At speaking engagements I have had across the country, numerous feminists have expressed indignation at the sexist views espoused by anthropologist Lionel Tiger. Close study of his best-known work, *Men in Groups,* shows ample cause for their anger.

Currently a professor at Rutgers University, Tiger came into prominence with the publication of this book in 1969, not long after the feminist movement surfaced. A male-biased viewpoint in the social sciences is not new. But Tiger's two-faced stance puts him in a special niche. Barely concealing his animosity toward women, he poses as their benevolent friend. His is an ultrasophisticated and crafty line of justification for male supremacy.

Tiger's central thesis is that only males have the capacity to form attachments to one another, a trait he calls "male bonding." He declares that this is built into the "biological infrastructure" of the male sex alone and has no counterpart among females. A female can form maternal bonds with her offspring and a sexual bond with the male who impregnated her, but, according to Tiger, she is incapable of forming any bonds with other females.

Tiger attributes this superior trait of males to the fact that they are the hunters while women are only the breeders. Since men hunt animals "co-operatively," they alone can form the close-knit bonds stemming from this occupation. This is the basis for the admiration and love found among "men in groups." Since women are not hunters, they do not and cannot inspire such solidarity, trust, and affection among themselves. Hence women do not form groups. Moreover, male bonding is a permanent feature of masculine relations while the sexual tie between a man and a woman is "ephemeral." Hunting and male bonding are at the bottom of eternal male supremacy. Women can do nothing about their deficiency because they have been shortchanged by Mother Nature in their biological infrastructure.

In this effort to make women inferior to men Tiger makes hunting the prerequisite for male bonding, beginning with the primates. This is in defiance of the fact that primates are vegetarian animals, not hunters of flesh foods. Some baboons occasionally eat the flesh of killed animals but this is exceptional; most primates in the wild eat no meat.

Tiger himself admits that "meat-eating constitutes a tiny proportion of baboon diet." According to his own figures they are 90 to 98 percent vegetarian. He further states that "meat-eating is learned behavior" (pp. 34-35). Thus hunting, on which his whole thesis hinges, began only with the advent of the hominids and their development of a carnivorous diet about a million years ago.

This does not prevent Tiger from claiming that hunting has been the decisive factor in prehuman primate evolution going back to fourteen or more million years ago. "The crux of my argument is that male bonding patterns reflect and arise out of man's history as a hunter," he writes, and "man's major evolutionary specialization was an ability to hunt animals cooperatively." He goes on: "So, in the hunting situation, it was the hunting group—male-plus-male-plus-male—which ensured the survival of the entire reproductive community. Thus was the male-male bond as important for hunting purposes as the male-female bond was for reproductive purposes, and this is the basis of the division of labor by sex" (pp. 122, 126).

Unfortunately for Tiger's thesis, in the animal world it is not the males but the females who possess the rudiments of cooperation or "bonding." This came about as a result of their maternal functions and the provision and protection they furnish to their offspring. An extension of these maternal functions makes it possible for animal females to band together in maternal broods, herds, or packs. Males, on the other hand, in their competitive striving for dominance, one against the other, do not naturally possess this rudimentary form of cooperation. Solidarity among males is a cultivated trait, learned under human and social conditions. Thus Tiger turns the real situation upside down. Animal "bonding" exists not in the male but the female "biological infrastructure."

To make his case for male superiority Tiger resorts to a distorted anthropology as well as a falsified biology. There have been two major turning points in human evolution: the first a million years ago when our primate progenitors transformed themselves into hominids, and the second a few thousand years ago when primitive matriarchal collectivism gave way to patriarchal class society. Tiger erases both of these qualitative leaps in human development.

Genuine scholars draw a clear distinction between organic or biological evolution and the social/cultural evolution that emerged with human life. The eminent paleontologist George Gaylord Simpson emphasized in *The Meaning of Evolution* that human evolution was of a different *kind* than purely animal evolution. The biological factors that dominate animal life have been largely displaced in the human world by social factors and cultural conditioning.

Tiger rejects these decisive determinants of human life and tries to downgrade scrupulous scientists who acknowledge them. In his typically devious way he writes that "socialists and other reformers maintained that human behavior differed in kind from animal behavior" (p. 11). For Tiger all evolution is of one kind—biological. Man is essentially an ape—just another primate—with woman presumably standing even lower, a kind of sub-ape.

The fallacy that man is governed by the same behavior patterns as animals is known as "social Darwinism." This theory came into existence in the late nineteenth century as a justification for unbridled capitalist competition. Subsequently, because of its gross distortions of both human and animal behavior, it fell into disrepute. Tiger himself refers to the "unfavorable reception" accorded social Darwinism. At the same time, by reducing social to biological evolution, and man to primate status, Tiger shows his own affinity with this discredited theory. He belongs to the school of vulgar biologism that was made popular in the 1960s by Robert Ardrey, Konrad Lorenz, and Desmond Morris.

Not content with reducing humans to animals, Tiger seeks to elevate animals to human status. He thinks that animals are not only social but "cultured" beings. In all seriousness he acclaims "the discovery that primates develop cultural forms in addition to species-specific 'programmed' behavior" (p. 22). To him the special artifacts, achievements, institutions, and abilities of humans are as much a part of animal as of human life—even without the presence of any ape scientists, artists, writers, and anthropologists to prove it.

This anthropomorphic biology is accompanied by a completely unhistorical anthropology/sociology. To Tiger all societies from the most primitive to the present have been patriarchal. He refuses to recognize the matriarchal epoch when, although social and sexual equality prevailed, women were the leading sex. He dismisses the matriarchy as representing no more than "matrilineal descent patterns." Yet he fails to explain how "the principle of hereditary succession" through the female line could come into existence in the first place if society has always been patriarchal and male-dominated (p. 92).

Such is the rickety methodological structure on which Tiger bases his assertions of female inferiority and his glorification of male supremacy. Let us examine his position in more detail.

On Female Inferiority

When Tiger speaks of "men in groups," he is not referring to the huge groups of men, organized and unorganized, that comprise the massive modern working class. If he did so, he would be obliged to refer as well to the "women in groups" whose steadily increasing numbers have almost reached the halfway mark in the same work force. This would hardly fit in with his thesis that women, "alas," are incapable of grouping because they do not possess the biological infrastructure of the male bonders. And it would give the lie to his charge that women are so dependent upon men for financial support, for sex, and as reproductive partners that they easily become "strike-breakers" against other women (p. 272).

Tiger gets around this difficulty of very visible masses of women grouping and fighting together in labor struggles by downplaying the whole period of capitalism. It is only about two hundred years old anyway, he observes, which is insignificant compared with the many "millions" of years of hunting activities.

Thus when Tiger speaks of "men in groups," he is referring to the resplendent figure of Man the Hunter. "Hunting *is* the master pattern of the human species," he writes. "Most humans at most times—with the exception of several thousand recent years—have needfully existed with awareness of their symbiosis with animals" (p. 216, emphasis in original).

Tiger the anthropologist shiftily sidesteps the fact that the hunting-gathering epoch began to go out of existence some eight thousand years ago, at which time hunting as a "needful" occupation receded and was finally abandoned. With the advent

Women "bonding" with women (above). A convention of the Coalition of Labor Union Women (CLUW). Women "bonding" with men (below). A strike action of teachers.

of agriculture and stock-raising, men moved up to the higher occupations of farming, herding, and crafts. A new "symbiosis" with animals then took place as hunting animals for food was reduced to a sport.

But Tiger believes that once a hunter, always a hunter—a corollary of his view that once a primate, always a primate. Both fit in neatly with his central thesis of Man the Hunter—the Superior Animal, and woman the breeder—the inferior creature.

In developing this theme Tiger offers a few embellishments of his own to the standard male-biased line that women are handicapped by their childbearing functions and other disabilities that make them economically dependent upon men and rivet them to nursery and kitchen. To this Tiger adds that women lack the striving for dominance and the aggressiveness that endow the male sex with the superior physique and intelligence required for work, play, and politics. To substantiate his point, Tiger refers to the "sexual dimorphism" found among primates—the set of normal differences between the sexes. Selecting certain monkeys as models, he writes that "both baboon and macaque males are very aggressive and intensely concerned with dominance." On the other hand, female primates do little more than "minor bickering," without attacking or killing one another (pp. 35, 37). This "dimorphism" separates the men from the girls.

It is true that one of the most pronounced differences between the sexes in the animal world is the aggressive-dominance trait of the males, not simply among baboons and other primates but among all mammalian species. However, this is not so desirable a trait as Tiger represents it to be. Under nature's blind rule, males are antagonistic to other males, competing with one another for the dominant place in the female group. Since only a few can win, the rest become peripheral and expendable. Some are killed in the struggles, others become "loners" trying to find a group to which they can attach themselves.

A Learned Trait

This trait of the male sex was overcome only after our branch of the higher apes passed over into human life. Through their newly acquired labor and cultural activities men learned how to hunt together instead of aggressively ousting one another. Thus the "sexual dimorphism" among animals favors not the males but the females; for it is the females who possess the natural trait of cooperation. Male bonding is a learned trait that was acquired by the male sex in the transition from ape to human. And it was learned from the female sex.

One of the most important tasks that confronted our female ancestors was to moderate and suppress the aggressive-dominance impulses of the males and transform them from antagonistic, competitive animals into a cooperative brotherhood of men. The anthropological record shows how this was accomplished. Through the social regulations laid down by the women, the "fratriarchy" arose as the male counterpart of the matriarchy. This matriarchal-brotherhood was an egalitarian, collectivist society as beneficial for men as it was for the women who took the lead in its creation.

Far from dominating women and degrading them, primitive men held women in the highest respect and esteem. It was not until the collective clan-brothers of the matriarchal system were overthrown by the propertied fathers of the patriarchal system that women became the dominated and degraded sex.

In Tiger's mechanical view of the primitive division of labor by sex, men were the hunter-workers and women merely the procreators of children. However, the productive role of primitive women was not minor, as he implies, but major. They were not only the procreators of life but also the chief providers of the necessities and comforts of life.

While men were hunters, women were the food-gatherers and cultivators of the soil; they were the cooks and preservers and storers of food for future use. Their industries included all the crafts from basketry and leather-making to pot-making and architecture, etc. In the course of their work they developed the rudiments of science, medicine, art, and language. They domesticated plants and animals and built the settlements without which cultural life could not have existed. They were the first ambassadors and peace-makers. (See *Woman's Evolution*, pp. 105-24.) All this is unambiguous evidence of the priority of the matriarchy.

Tiger seems to be well aware of this evidence. But in typically dishonest fashion he doesn't openly argue the issues. He refers vaguely to "a carping folklore about 'woman's logic'" (not her work!) and irrelevantly remarks that men are not as "fickle" as women. Then he adds: "I hesitate to enter a controversy about the comparative skills of males and females. Nor shall I comment here on the felicitous or unfortunate effects of greater female participation in momentous matters of communal life. It is a controversy, anyway, which eludes theoretical solution" (p. 113).

After this cavalier dismissal of women's productive and cultural record, Tiger presents his own theories about women's eternal inferiority. Some of these are downright silly. For example, women "are less able to endure heat than males," which "was a disadvantage to women hunting in tropical countries of southern Africa" where the behavioral patterns of *Homo sapiens* were formed. Again, there are differences "in the sexual response cycles of males and females" and in the "development of gender identification" that make men and not women the hunters (pp. 123-24). Elsewhere he has the following gem: "females throw missiles, spears, etc., with much the same motion primates use," unlike males who throw them in the superior manner of human hunters (p. 144).

Hormones to Genetics

If arguments like these are unconvincing, Tiger has more. He leaps about like an agile monkey, from hormones to genetics, to explain the hazards posed for the human species should women engage in hunting. The specialization of males for hunting, he writes, "favored those 'genetic packages' which arranged matters so that males hunted co-operatively in groups while females engaged in maternal and some gathering activity." Any disturbance of this male specialization would have been disastrous to the "genetic packages" that arranged matters this way, and presumably nothing could disturb them more than females taking up hunting.

Moreover, on the other side of the question, females who hunted would be displaying "nonmaternal female behavior," and that would be equally bad for the "genetic pool." The latter is safe only with "those females who accepted a clear-cut sexual difference and enhanced the group's survival chances chiefly by full-time maternal and gathering behavior" (pp. 57-59).

All this absurd advice to women is buttressed by similar advice to men—at least those who might be softheaded about letting women hunt. According to Tiger, this "could interfere with the co-operative nature of the group by stimulating competition for sexual access." Male bonding seems to rest upon fragile foundations after all—despite Tiger's assertion that it has been embedded for millions of years in the male biological infrastructure.

In any case, there is an even worse hazard than this sexual

competition among males for access to females. Tiger writes that "males who accepted females into their hunting groups would be, like those females themselves, less likely to add to the genetic pool than presumably more efficient hunters who maintained male exclusiveness at these times" (pp. 59-60).

If these arguments remain unconvincing, Tiger resorts to even more vicious devices for intimidating women into accepting his doctrine of exclusion. He writes, "Those females who hunted with males could not reproduce as numerously as non-hunting females," and he points scornfully to "the figure of the dry spinster of Euro-American cultures or the barren woman of Africa and Asia."

He reserves his most horrible examples for North American women. He writes, "There is a contemporary version of the 'genetic impotence' of 'hunting women'—I refer to females involved in relatively high-influence, high-status work. For example, among female executives in North America an unusually high proportion, about a third, are unmarried, while those who are married tend to have no children or a number lower than the average" (p. 127). This is definitely not the kind of women "preferred by men." According to Tiger, while men certainly prefer lovely, talented, and wealthy women, they must above all be "benign forces to involve in the homes of men" (p. 182).

Tiger's flimflam on the undesirability of allowing women to hunt—an occupation that has long been extinct—is only a cover-up for his rejection of women working at "men's" jobs or engaging in politics today. He is explicit on the point that woman's place is in in the home and not in the social arena sharing decision-making powers with men.

"The public forum is a male forum" in which "females do not participate," he says. Today "it is men who dominate the public and private State Councils of the world; men are the Ambassadors, the Linguists . . . there are few 'Spokeswomen' "(p. 75). He dismisses the argument that this is due not to any biological disabilities of the female sex but to an oppressive society that deliberately excludes them from public life.

To Tiger politics is war and men are the hunters and warriors—the capable, intelligent, war-making sex, and the pillars of social life. He writes that "human nature" is such that it is "unnatural" for "females to engage in defense, police, and, by implication, high politics" (p. 112). Even in work he says that men can do all kinds of jobs, but not women. "Why cannot females become fighter pilots, tank commanders, or police chiefs? When men join armies they become dishwashers, laundrymen, and nurses. Yet when women join armies they do not commonly take jobs which are conventionally defined as masculine" (p. 108).

Perhaps by now, some seven years after he wrote this, Tiger has learned that it is social discrimination and male prejudice that have held women back from taking men's jobs. In the few years that the feminist movement has been on the march, women have already broken down many barriers; they have moved into scores of occupations formerly reserved for men only.

Today there are women architects and sportscasters, bartenders and engineers, doctors and train operators, stevedores, miners, and truck and bus drivers. Women are rig operators, coast guards, and construction workers, as well as fire fighters, pilots, and subway operators. They are electricians, ditchdiggers, and cops. Women have broken down their exclusion from Olympic sports as well as from the major male-supremacist universities. They are now being admitted into the military academies of West Point, Annapolis, and Colorado Springs. If they are not yet tank commanders, bomber pilots, or astronauts, this is not because of any feminine frailties.

The handicaps placed upon women are exclusively social and not biological; for the earliest and longest period in human history there were no such handicaps. Tiger's notions about the innate inferiority of women are as outdated as the horse and buggy.

On Male Supremacy

Tiger does not regard production of the necessities of life as the unique attribute of humans—as the activity that elevated them out of the animal world. To do so would involve disclosing the prime part played by the women. He begins with the capacity of men to male-bond and writes: "Male bonding I see as the spinal column of a community, in this sense: from a hierarchal linkage of significant males, communities derive their intra-dependence, their structure, their social coherence, and in good part their continuity through the past to the future" (p. 78).

This does not answer the key question: how did the aggressive-dominance trait of male animals that leads to antagonism and isolation give way to the human capacity of men to unite through cooperative bonding? Tiger's answer is somewhat astonishing.

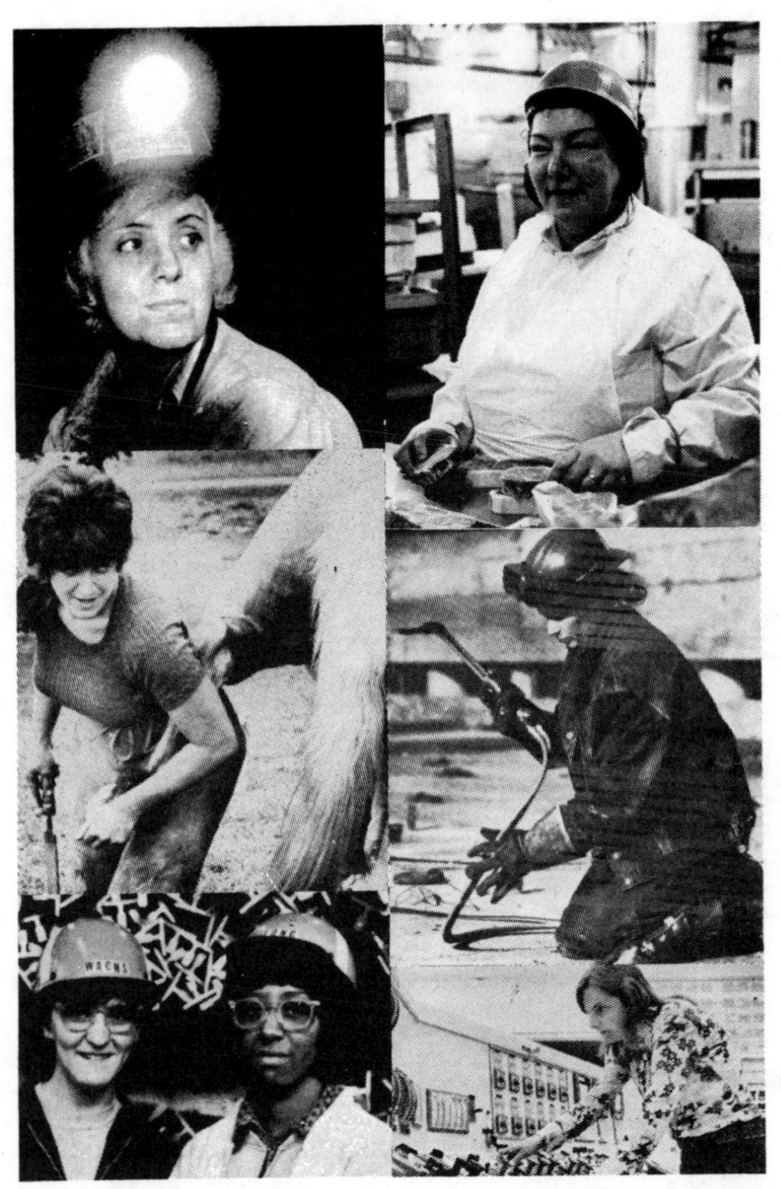

Women working "men's jobs." Clockwise from top right: butcher; welder; control-room operator in electric power station; steel workers; blacksmith; miner.

Competitive aggression and cooperative bonding are one and the same trait! He writes in all seriousness that "aggression is an intensely co-operative process—it is both the product and the cause of strong effective ties between men" (p. 247).

However, it seems that these "strong effective ties" do not transcend class or racial barriers. A man, says Tiger, will not bond with just anybody; "he will bond with particular individuals because he has certain prejudices and standards in terms of which he is willing to bond" (p. 28). Thus "males will prefer to be with high-status males. . . . They gain status themselves from the positions of their companions. A group of men conscious of its status, power, and security differs in emotional tone from a group in the Bowery" (p. 184). In short, high-class men preserve their snobbish separation from low-class men.

The same separateness exists between white and Black men. Tiger writes (parenthetically) that "there will be some who, for various reasons, prefer culturally unpreferred persons." The example he offers is not that of an upper-class man who broke the racial barrier but of a dismal female: "The rich white girl and her poor Negro chauffeur of Richard Wright's *Native Son* is a case in point." He also refers to many of William Faulkner's stories, which "revolve about a similar deliberate violation of socio-sexual norms" (p. 184). Tiger's male bonding is strictly class and race stratified.

Within this elitist framework Tiger gives the "logic" of his argument: that "males are prone to bond, male bonds are prone to aggress, therefore aggression is a predictable feature of human groups of males." He is opposed to changing this feature. He writes, "To reduce opportunities for such aggression is to tamper with an ancient and central pattern of human behavior" (p. 241). In Tiger's view, since animals are killers, human males are bound by the same inherited behavior pattern.

For men who might be repelled by this thesis, Tiger issues a warning: they place themselves in the same impotent position as infertile women. He writes

> bondless, aggressionless males are in a real sense equivalent to childless females. Of course, childless females are viable and many choose their condition and enjoy its benefits. At the same time [they] may be held to have lacked participation in a massive biological activity and its psycho-social consequences. In the same way, friendless inhibited males are not only friendless and aggression-inhibiting, but possibly do not experience the male equivalent of child

reproduction, which is related to work, defense, politics, and perhaps even the violent mastery and destruction of others [p. 242].

Glorification of War

This "bonding-cum-aggression," as Tiger dubs it, clears the road for his glorification of the masculine traits of war and violence. War is almost universally an all-male enterprise, he writes, and "other agencies of aggressive-violent mastery are composed of males" (p. 226). Among these agencies he cites the Nazis and the Ku Klux Klan. In passing, he notes the few unspectacular things women can organize: "from charity groups to hairdressing salons to berry-gathering cliques" (p. 218).

The exceptional merit of male bonding-cum-violence is that women are excluded. He writes that "males bond in a variety of situations" which involve power and force, and they "consciously and emotionally *exclude* females from these bonds" (p. 143, his emphasis). One well-known example, cited by Tiger, is that of the Hitler era in Germany, when Goebbels elaborated the doctrine that woman's place was in the kitchen and nursery—"kinder, kirche, und küchen." He offers this as further evidence of the innate inferiority of the female sex and writes, "apparently even major changes in political form and ideology can have little effect on the role of women" (p. 90).

To Tiger fascism is merely a change in "ideology" and not a disastrous political defeat of women and workers. This is in keeping with his thoroughly reactionary and racist outlook which sets whites above Blacks and men against women, and which glorifies wars and other forms of the "violent mastery of others" as the highest virtues of Manhood.

On this basis Tiger has harsh words for women warriors—they are as bad as women hunters. In his opinion women who try to join armies are disparaged as "necessarily often transvestite" (p. 104). On the other hand, Tiger is favorably disposed to homosexuality—providing it is male—and he proposes separate men's houses and all-male societies and clubs where husbands can find enjoyment and relaxation apart from the tedium of wives and domesticity.

When Tiger's book appeared, provoking expressions of outrage, the *New York Times Magazine* generously accorded him a forum in which to defend himself. His article appeared on October 25, 1970, under the caption: "Male Dominance? Yes, Alas. A Sexist

Plot? No." This disclaimer did not alter his image as a gross upholder of male supremacy. Rather, it reinforced his book, which, despite its shiftiness and sophisticated pseudoscientific terminology, is essentially the self-portrait of a virulent woman-hater.

Anthropology and Feminism
An Exchange of Views
(1975)

HAYMES'S CRITICISM OF ENGELS AND REED

The more orthodox Marxists in the women's liberation movement and other feminist radicals often enough conduct their arguments within the context of nineteenth-century anthropological speculation. This is somewhat of a paradox. The changes wrought by events of the twentieth century—greater educational and vocational opportunities for women, the "home front" activities of the two world wars, development of better contraceptive techniques, more labor-saving devices for the housewife, etc.—presumably provided the backdrop to feminism in this century. How could a branch of the social sciences developing in Victorian times provide a useful ideological prop for current feminism? How might a view of women growing out of the thinking of a period one hundred years ago provide a useful argument for the rhetorical arsenal of modern feminists?

Anthropologists such as Edward B. Tylor, Herbert Spencer, James Frazer, and Lewis H. Morgan dominated the newly emergent social science of the last half of the nineteenth century. Darwin's influence on early anthropology is generally accepted as seminal; hence their thinking was "evolutionist."

As a prime example, Lewis H. Morgan's work combined evolutionary thinking with the concept of "survivals." He studied North American Indians firsthand and generalized from his knowledge of kinship relationships to argue for a theory of

---

This exchange between Dr. Howard Haymes and Evelyn Reed appeared in the March 1975 *International Socialist Review*. Both contributions were written before the publication of Reed's book *Woman's Evolution*. Haymes holds a doctorate in education from New York University and is presently the director of Continuing Professional Education in the School of Nursing, Health Sciences Center, at the State University of New York at Stony Brook. He has contributed articles to the *Journal of the University Film Association* and *Psychiatry*, and is author of *Feature Films in the Classroom*.

"primitive promiscuity." Promiscuity, however, was not observed; it was inferred. This inference was made by such reasoning as follows:

Morgan found that among the Seneca-Iroquois, the term for my father's brother is "my father"; both are addressed as *Ha-nih.* The term for my mother's brother is "my uncle" or *Hoc-no-seh.* In the same way, assuming one to be a male, his brother's son is "his son," or *Ha-ah-wuk.* But his sister's son is "my nephew," or *Ha-ua-wan-da.* If one were female the situation would be reversed. These facts struck Morgan as apparently illogical, because people related to a given son as "uncle" or "cousin" who stood in the same "degree of nearness" but were linguistically separate. This constituted a "quaint survival." Part of the language changed as the conditions of Iroquois civilization changed, but part did not. From this, a clue, as Morgan saw it, was obtained. Specifically, sometime in its more primitive past, group marriage had been a practice of the Iroquois. Only linguistic evidence remained. Other living, "primitive" societies showed similar patterns too; the patterns were not restricted to the Seneca-Iroquois.

That "sexual promiscuity was the first conceivable stage of human society," was the only inference Morgan could draw from his evidence. So wrote Kardiner and Preble in *They Studied Man.* The strictures against incest were seen "as a reformatory movement to break up the intermarriage of blood relations." Thus Morgan felt he had the key to exogamy (marriage outside the group).

In *Totemism and Exogamy,* another evolutionist, James Frazer, proclaimed that exogamous group marriages "gradually evolved into individual marriages" (cited by Kardiner and Preble, p. 91).

Also among the earliest anthropological treatises in the evolutionist tradition was the work of Johann Jakob Bachofen. *Das Mutterrecht* was published in 1861. In this book, "The Mother-Right" or "Matriarchy," Bachofen claimed to have unearthed the existence of a universal stage of human development—promiscuity. From this first stage:

> . . . a family group, strictly organized by the mother and consisting of her children without regard to paternity [developed] graduating into a matriarchally controlled family with acknowledged fathers, and replaced finally by the patriarchally controlled family so familiar to the Europe of Bachofen's day [cited by Helen Diner, in *Mothers and Amazons*].

Professor James L. Gibbs, Jr., in 1964, assessed the work of these early evolutionists:

> It [matriliny] was found, they suggested, in societies in which the biological role of the father was unknown or where mating was so promiscuous that paternity would necessarily be in doubt. . . . matriliny would lead to matriarchy or the concentration of economic, political and ritual power in the hands of women. In their unilineal view of human history, matriliny would lead . . . to patriliny, supposedly a higher form of family organization [*Horizons of Anthropology,* pp. 161-62].

However, by 1890 the evolutionists had come into disrepute within the anthropological profession. Emile Durkheim, Franz Boas, and later Bronislaw Malinowsky, Ruth Benedict, and A. L. Kroeber took anthropology away from evolution and toward functionalism, which was "less speculative" and had more precise methodology.

The mainstream of anthropology was no longer to be concerned with evolutionary theories, matriarchies or "survivals." Marx and Engels, however, were influenced by evolutionist thinking and they and their later followers kept the tradition alive. In the opinion of this writer the functionalist objections of Boas and Durkheim are easier to accept than the unilinear view of Morgan, Engels, and, for that matter, the Soviet anthropological community today. Unilinear evolutionary doctrine *assumes,* but does not prove, that cultural characteristics are necessarily derived from identical causal processes. The variability of a given possible cultural response has been underrated by the partisans of the unilinear tradition. Furthermore, Morgan himself, a field worker of note, had too few ethnographic field studies to rely upon in reaching conclusions about the ultimate diachronic question (i.e., inevitable stages of historical development).

Evolutionist reasoning from nineteenth-century anthropology was downgraded and discarded by the inheritors of the traditions of Durkheim and Boas, the modern anthropologists. The unilinear concept of human universalist progress survived among the heirs of Marx and Engels.

The works of Bachofen, Tylor, and Morgan were not available to the authors of *The Communist Manifesto* in 1848, but later Engels did incorporate evolutionist thinking about the primeval "universal" matriarchy of primitive man. By comparing the *Manifesto* with *Origin of the Family, Private Property, and the*

State, it can be seen that evolutionist anthropological thinking has been adopted by Engels. Whereas the *Manifesto* contained an analysis of women in bourgeois society, the latter work outlines the unilinear view of the development of the family along lines that Bachofen, Tylor, and Morgan suggested.

Returning to prehistory, Engels claims that the pairing marriage of barbarism gave way to monogamy because of the creation of a surplus. With the existence of a surplus, man gave up hunting and took up permanent residence in the household. Women and men shared equally. As the surplus grew, woman, who was once the chief creator of wealth by virtue of her domestication of animals and plants, began to take second place. Her production was confined to the home. Men increased their wealth and later took slaves to provide even more individual, single-family wealth. The slave, taken by war, thus became the first oppressed group. Woman, cut off from social production, became a mere breeder of inheritors of private wealth. Patriliny was now established; mother right was abolished and with it "the world historical defeat of the female sex." This now marks the beginning of the historical era, i.e., ancient society.

Evolutionism and Modern Feminism

The modern feminists have mixed nineteenth-century evolutionist anthropology with the perspectives of Marx and Engels on marriage and family. A theoretical morass was created. Some of the feminists borrow analyses taken over by the Marxists, who in turn borrowed from Morgan and Bachofen. Some women's liberationists have returned to the original sources. Some produce a mixture of Marxism and anthropological "evolutionism." Other feminists, such as Ti-Grace Atkinson, create highly individualistic, dialectical theories of female oppression related, but only vaguely so, to classical Marxist interpretations. With writers like Mary Jane Sherfey and Elaine Morgan, economic theory and old anthropology merge with biological theories constructed in the light of the latest research on sexuality.

Self-proclaimed Marxists like Evelyn Reed and Juliet Mitchell call specifically for the overthrow of capitalism, not the mere abolition of abortion laws or the passage of constitutional amendments. These so-called politicos view the women's movement as only a portion of a larger historical dialectic that will culminate in a classless, stateless, oppression-free utopia.

Other feminists have bolted from leftist-oriented movements

and have instead opted for a "women-first" stance. In doing so they are at odds with the so-called movement men whom they often roundly denounce, as well as the "politicos" like Reed and Mitchell. Robin Morgan, Marlene Dixon, Marge Piercy, and Pat Mainardi, to mention only a few, are on record as alienated from men in the leftist organizations. The same is true for Beverly Jones, an SDS "dropout." The stereotype of the "sexist New Left male" almost becomes a kind of inverse litany with a quote attributed to Stokely Carmichael by no less than five authors: "The only position for women in SNCC is prone."

Much of the tension between the leftist movements' "politicos" and the "feminist-first" groups centers about differing conceptions of women as a separate "class." For example, Kate Millett claims that women only apparently, in a superficial way, belong to different socioeconomic classes. Roxanne Dunbar sees women as belonging to an inferior "caste." Evelyn Reed offers the "correct" Marxist interpretation that women are an "oppressed sex." Germaine Greer, commenting on Reed's work, says her arguments "are couched in typical Marxist doctrinaire terminology, buttressed by phony anthropology and poor scholarship."

The capitalist system comes under a veritable verbal bombardment at the hands of the women's liberationists reviewed in this study. Even the older, more "conservative" writers like Bird and Friedan see American capitalism as "exploiting" women. Authors like Morgan, Firestone, Greer, Atkinson, Reed, and Dixon call for truly radical (i.e., comprehensive) changes in the economy. Marx's and Engels's interpretations of capitalism and clearly collectivist solutions are freely and frequently mentioned. Here are some readily obtained examples from a seemingly inexhaustible list:

Morgan. ". . . the profoundly radical analysis beginning to emerge from revolutionary feminism [is] that capitalism, imperialism, and racism are symptoms of male supremacy-sexism" (*Sisterhood Is Powerful,* p. xxxiv).

Firestone. "Women today are still excluded from the vital power centers of human activity . . . women are like a Petty Bourgeoisie trying to open up shop in the age of Corporate Capitalism" (*Dialectic of Sex,* p. 166).

Millett. "Prostitution is, as Engels demonstrates, the natural product of traditional monogamous marriage" (*Sexual Politics,* p. 122).

Figes. "The rise of capitalism is the root cause of modern social and economic discrimination against women" (*Patriarchal Attitudes,* p. 65).

Sachs. "A woman's child-raising responsibilities are the excuse given by the capitalists for assigning women to a superexploited position" ("Social Bases for Sexual Equality," in *Sisterhood Is Powerful,* p. 462).

Reed. ". . . as the liberation movement acquires a stronger thrust and penetrates deeper among working women, it can act as a catalyst to set the anticapitalist potential of the working class forces into motion. . . . 'we have nothing to lose but our chains; we have a world to win'" (*Problems of Women's Liberation,* p. 63).

Dunbar. ". . . sexism is the opiate of the American people" (quoted in Bird, *Born Female,* p. 223).

Beyond merely the employment of leftist rhetoric, radical feminists have clearly adopted interpretations of women's roles based on Engels's analysis. Marlene Dixon, for example, writes: "The American system of capitalism depends for its survival on the consumption of vast amounts of socially wasteful goods, and a prime target for the unloading of this waste is the housewife" (*Ramparts,* November 1969, pp. 60-61).

Further on she writes: "Frederick Engels, now widely read in the women's liberation movement, argues that regardless of her status in larger society, within the . . . family the woman's relationship to the man is one of proletariat to bourgeoisie. One consequence of this . . . is to weaken the capacity of men and women oppressed by the society to struggle together against it."

Evelyn Reed weds Marxism and women's liberation with the warning ". . . though the full goal of women's liberation cannot be achieved short of the socialist revolution, this does not mean that the struggle to secure reforms must be postponed until then. It is imperative for Marxist women to fight shoulder to shoulder with all our embattled sisters . . . from now on" (*Problems of Women's Liberation,* p. 75).

Whether or not the women call themselves "Marxists" is not at issue. Some do; some do not. In broad terms, however, one may trace Marx-Engels's view of primitive, ancient, feudal, and capitalist societies, and then show how this analysis is often freely adapted by women's liberationists. In short, the philosophic structure of the feminists often rests on borrowed leftist theory. Other authors, however, concern themselves with sexism in a historical sense; they are concerned with both history and prehistory as clues to modern "sexism." They venture back to paleolithic and neolithic times; hence, their view is "anthropological."

The Reed Model

Reed assumes that the primeval division of labor was equitable. Production was "social," meaning "communally undertaken." Society was supposedly "oppression-free" since goods were held in common and a surplus, the "material basis for class exploitation," was at first nonexistent. Separate families were nonexistent; all were "social brothers and sisters." Children were the responsibility of all; the question of who one's father or mother might be was irrelevant.

Savage society could be deemed "not only a matriarchy, but a fratriarchy—a 'brotherhood' of men." All women were "mothers"; all men were "uncles" (mother's brothers) to the children. When "pairing families" arose, husbands simply replaced the mother's brother, but democracy and economic equality were retained.

With men away hunting, women gathered edible flora. Women also gathered insects and small animals for the collective diet. In time women developed agricultural techniques as an outgrowth of plant gathering. They also developed herding as an outgrowth of handling small animals.

At the point where a surplus arose, men gave up hunting except as sport. Agriculture and herding and the permanent fixing of the men within the settled community provided the basis for even more rapid progress. Savagery and civilization were then demarcated. Men began to take over agriculture and industry necessitating a redefinition of woman's role. A woman no longer performed social (communal) functions. She was biologically defined now as an individual man's wife and mother of his child. Her production was "private," i.e., she merely tended his house and did so without contact with other women. Economic leadership roles were ironically reversed.

This researcher has relied upon Reed's analysis mainly because it was the most carefully elaborated discussion available from all of the feminist writers studied. Reed is a Marxist of the "politico" variety (i.e., woman's problems are only part of a larger class problem). Despite the doctrinal haggling over such "niceties" as to whether women are a "caste" or a "class," or "feminism first" versus "socialism first," Reed's perspective is a useful model. Many of the other feminists subscribe to various parts of her analysis of the origin of "women's oppression" whether or not they approve of Reed's overall political stance.

With a view of feminist "anthropology" set down, Reed proceeds to look at modern marriage. For her, marriage today is the result of a dialectical process rooted in economics. The current mode is not fixed by biology. For instance, at one time, in matriarchal primitive society, no one was legally married.

Modern marriage serves several purposes. The first is as a means of transmitting property; this is the original (ancient historical) *raison d'etre.* The second purpose of marriage is to "relieve the capitalists of all social responsibility for the welfare of the workers and dump . . . burdens upon the poor in the form of family obligations." Another purpose of marriage is to construct the family to serve as a consuming unit. Reed writes "love becomes measured by the number of things parents buy for their children and do for them in the way of special privileges and pampering. This in turn makes the children *private possessions* . . . and places them under control like any other form of property" (*Problems of Women's Liberation,* p. 61, emphasis added).

Further, the nuclear family is described as "narrow," "embittered," and "ingrown." Disharmony is "typical."

While love is the theoretical ideal upon which marriage is founded, for Reed the realities of marriage are grim. Since most women must be dependent housewives, husbands are perforce mere "good providers." If wives work, they are "doubly oppressed." They are thus ". . . exploited on the job by their employers and oppressed at home through family private servitude."

The idea that love was the foundation of marriage was dismissed by Reed as a kind of myth. Feudal marriage (in the next Marxist stage of history) was simply a device for transmitting land. Romantic love was outside of marriage. Love and marriage were only fused at the beginning of capitalist society. But Reed writes of this merger:

> This does not mean that the combination was a smashing success since love had to contend with so many adverse factors. It is certainly true that among the working people . . . mutual attraction and love are normally the basis for marriage. But it is not true, as the fairy tale has it . . . that the couples live happily ever after. From the statistics we can see that the marriages of the working people are being battered and shattered as rapidly as those of the middle classes and the rich [ibid., p. 48].

Reed comments on another aspect of marriage: sexuality. Current feminism's view of sex is contrasted with the older women's rights movements. The nineteenth-century movement fought for legal rights, but stood clear of sexual matters. Now, she says, the "hypocritical 'double standard' which gave men sexual freedom, but denied it to women" is being discarded. Premarital and extramarital sex are now very common. So common, she writes, that marriage is collapsing; it is a dying institution.

Germaine Greer's writing about marriage, family, and sexuality follows similar lines. This appears true despite the fact that Greer denounced Reed as being "naive" and given to using "typical Marxist doctrinaire terminology" (*Female Eunuch,* pp. 317-18).

Greer's prescriptions appear somewhat similar to Reed's primitive "matriarchy" of prehistory: "Now that cloistering of wives is an impossibility, we might as well withdraw the guarantee [of paternity] and make the patriarchal family an impossibility by insisting on preserving the paternity of the whole group—all men are fathers to all children."

Greer also calls for a return to a communal society of women. She claims, as does Reed, that women are too isolated and engaged in "private" production. Greer writes: ". . . it would be a serious blow . . . if women shared, say, one washing machine between three families. . . They could form household cooperatives, sharing their work about, and liberating each other for days on end" (*Female Eunuch,* pp. 344-45).

Greer writes that "the family of the sixties is small, self-centered, self-contained and short-lived." The isolated young mother attempts to compensate for her lack of purpose by trying to "buy" her child's affection. As Reed said so Greer also says that family relations become exploitative:

> The working girl who marries, works for a period after her marriage and retires to breed, is hardly equipped for the isolation of the nuclear household. . . . Her horizon shrinks to the house, the shopping center and the telly. Her child is too much cared for. . . . The child learns to exploit his mother's accessibility, badgering her with questions and demands which are not of any real consequence to him, embarrassing her in public, blackmailing her into buying sweets and carrying him.

Other feminists use descriptions similar to Reed's. Karen Sachs is an example. In the following passage she relates "privatization" and capitalist profits: "For the capitalist, a society made up of nuclear families is a joy. It means lots of workers, lots of consumption units. . . . With small nuclear families, consumption is more inefficient" (in *Sisterhood is Powerful,* p. 462).

The reader will also recall Reed's view of the "family-as-consuming-unit," and "family-members-as-living-possessions" when Millett writes: "In noting its economic character, Engels is calling attention to the fact that the family is actually a financial unit. . . . Due to the nature of its origins, the family is committed to the idea of property in persons and in goods" (*Sexual Politics,* p. 124).

The final solution:

". . . the family, as that term is presently understood, must go. In view of the institution's history, this is a kind fate. Engels was heresy in his age. These many decades after, he is heresy still. But revolution is always heresy, perhaps sexual revolution most of all" (ibid., p. 127).

In short, the "new feminism" contains elements of some rather "old" theory. The anthropological speculations of the evolutionist school may have been generally disavowed by twentieth-century Western universities, but the unilinear view of societal development remained part of Marx-Engels's rhetoric. The result has been a curious blend of theory this writer has called "feminist anthropology."

Of course, not all feminists are prone to this kind of rhetoric. Writers like Betty Friedan, dubbed a representative of the "NAACP wing" of the movement by her feminist detractors, has opted for such mundane demands as equal pay for equal work, equal access to the professions, and equal opportunity for advancement within these professions. Such concerns seem "lackluster" and "run-of-the-mill' when placed beside the new evolutionism in the "grand old manner" of writers like psychiatrist Mary Jane Sherfey or Marxist-feminist Evelyn Reed. Feminist "anthropology," however, probably stands as good a chance of acceptance among most American women as evolutionism does among modern anthropologists—that is to say, small chance indeed.

In this author's opinion a more realistic perspective for women's organizations in seeking an end to antifemale discrimination would lie in a different direction. There are essentially five broad (and overlapping) areas of struggle for women's groups to do battle in so as to effect sexism's eradication: legal, educational, artistic, scientific, and ideological spheres. The ideological sphere is easiest in a sense.

What feminism seeks is equality of opportunity, one of the key premises of Locke, Jefferson, and the philosophers of the French revolution. Americans have an explicit commitment to this ideal. Freudian sexual maturational theory probably represented the last "respectable" bulwark against the granting of equal opportunity. Freud's whole doctrine of "femininity," from the penis-envy complex to woman's receptivity and passivity, is unraveling. Ideology and "science" offer, therefore, no barriers to equality of opportunity. The role of women's groups in concentrating on the three remaining arenas seems clear.

The states' departments of education must be pressured into an active program of nonsexist education. This task will be rendered substantially less difficult if the Equal Rights Amendment supporters find the remaining states necessary for ratification. In a stroke of the pen, as it were, all legal barriers would be erased. The last arena of the struggle for women's groups, the artistic, requires a concerted, sustained effort to enlist the support of creative people who can emotionally dramatize the inequities of past and present sexism. A massive consciousness-raising task is called for.

The perspective above excludes the socialist perspective defended by Reed. The "Reed model," it appears, represents the only women's liberationist view that is consistently "doctrinally pure." The other self-described leftist writers all seem to make the obvious error of forgetting the Marxist imperative to raise "class consciousness" as a prelude to comprehensive change. Although, unlike Reed, they have not "reconciled" feminism and Marxism, I do not share Reed's fear for their doctrinal "purity." This is so simply because of faith in the reform of the free-market system to eradicate sexism via means elaborated above.

I do not subscribe to the "inevitability" of the Marx-Engels unilinear view of history. I do feel that governmental amelioration of the vicissitudes of the marketplace is desirable in the direction of greater regulation and stimulation of the economy when necessary. Ultimately, the capacity of the American economy to produce unparalleled wealth, it seems to this writer, speaks more eloquently in defense of a basically free-market economy despite cyclical lapses. In summary, the prospect of a "unilinear leap" into collectivism is not desirable and certainly not inevitable.

EVELYN REED'S REPLY

Liberals and other fair-minded men do not object to the feminist movement—providing women restrict themselves to activities on such practical issues as equal rights, conduct themselves in a ladylike manner, and above all avoid controversial theorizing on the source of female oppression and its solution. Howard Haymes is one of these well-intentioned men who has the best interests of the women's movement at heart and gives his advice accordingly.

He surveys the galaxy of feminist writers who have articulated the problems and prospects for liberation over the past few years and finds cause for dissatisfaction. The women are too frank in their criticisms of sexist men and politics, and even of other feminists with whom they disagree. They engage in strident debates among themselves on the roots of male supremacy and female inferiority without consulting any male advisers and with no visible loss of solidarity as feminists. He deplores the "alienation" of New Left women from their men because the blemish of male chauvinism persists among them. But he is most distressed by the widespread influence of Engels and his Marxist followers upon the thinking of the movement.

Not only radical feminists but even more conservative writers, Haymes complains, have adopted Engels's thesis that women are exploited and oppressed by capitalist society. Whether or not they call themselves Marxists, the result is a "veritable verbal bombardment" of capitalism at the hands of these women liberationists. Moreover, he warns, this is not mere leftist rhetoric; it involves serious consideration of the socialist solution to fully eradicate the oppression of women. As Haymes puts it, "clearly collectivist solutions are freely and frequently mentioned" by the feminists.

No less disturbing to Haymes is the mounting feminist interest in anthropology in order to learn about the position of women in primitive society before they became the oppressed sex. According to Morgan and other evolutionary founders of the science in the last century, ancient society was matriarchal and collectivist. Women, far from being inferior, played a leadership role and held an esteemed position. These findings, and the full conclusions to be drawn from them, were set forth by Engels in his book *Origin of the Family, Private Property, and the State.* He showed how anthropological data confirmed the Marxist theory on the class roots of female oppression from slavery through feudalism to contemporary capitalism. Haymes feels called upon to refute the fusion of evolutionary anthropology with Marxist sociology in this classical work.

Opposition to Marxism in politics and to evolutionism in anthropology are not new; they have been rife ever since the two branches of science were born in the last century. What is new in the present situation is the women's upsurge, barely six years old, and the impact these two sciences have already had upon this expanding movement. Women are reopening and reinvestigating social and historical questions that have long been declared by academic authorities—mostly male—as conclusively settled and no longer debatable. Haymes wants to turn women away from this misguided course.

The mistake they are making, he says, is to mix together a "borrowed" leftist theory with a "borrowed" evolutionist anthropology. To this he opposes his own mixture, borrowed from anti-Marxists and antievolutionists. According to Haymes, socialism is no solution to the problem of women's liberation since a "classless, stateless, oppression-free" society is a utopia that can never be realized. He counsels women against accepting the findings of the pioneer anthropologists, who discovered a primitive form of classless, stateless, oppression-free "matriarchal" social order. This is incorrect, he says, since it has been declared wrong by the recent functionalist opponents of the evolutionists.

Engels himself made a mistake, in Haymes's view, by supporting these early anthropologists. How, he asks, can a nineteenth-century science be of any use to feminists a hundred years later? Today only "doctrinaire" followers of Engels, like myself, are keeping alive the discarded work of those "Victorian" anthropologists. The "Reed model," as Haymes calls it, is an unholy combination of feminism, Engelsism, and evolutionism. This witches' brew has had an adverse effect upon the theory and orientation of the women's movement.

Haymes refers to my writings to show how the new phenomenon of "feminist anthropology" has produced a "theoretical morass." Reed "assumes," he says, that primitive society was "oppression-free" and that all were "brothers and sisters." It was not only a matriarchy but—perish the thought—a fratriarchy, a "brotherhood of men." This kind of feminist anthropology, he opines, "stands as good a chance of acceptance among most American women as evolutionism does among modern anthropologists—that is to say, small chance indeed."

However, the women that Haymes speaks for with such assurance may have a few queries for him before they adopt him as their spokesman. Where does he stand on the question of the source of female inferiority today? If it is not attributable to a capitalist society that profits from the exploitation of women, does it come from the biological handicaps suffered by the female sex, as most men think? Haymes does not give his position on this crucial question.

Similarly, if Haymes rules out the discovery that women occupied a leading position in primitive matriarchal society, does he then agree with those modern anthropologists who say that because of their biological handicaps women have always been dominated by the superior male sex? If so, where is his evidence? So far he merely supports an unproven assertion.

In fact, what Haymes disqualifies as "feminist anthropology" is a serious effort on the part of women in the liberation movement to recover their own ancient history, which up to now has been closed to their inspection and knowledge. This has unavoidably reopened a long-standing dispute in anthropology between the evolutionists and their opponents. Since this is not simply an anthropological but, even more fundamentally, a political struggle, let us examine the politics behind the opposition to evolutionism.

Politics and the Flight from Evolutionism

The sharpest dividing line between the contending schools of anthropology is over the question of method. Morgan, Tylor, and other nineteenth-century scholars regarded their science as a study of prehistoric society from its beginnings up to the civilized period. Whatever their shortcomings and mistakes, which are inevitable in any new science, they made brilliant beginnings in reconstructing our most ancient history through their application of the evolutionary method.

Their twentieth-century successors, however, rejected the evolutionary method and changed the definition and goal of anthropology. Separate studies of diverse primitive cultures and peoples became an end in themselves, without reference to the clues they provided for reconstructing a connected chain of human development from one social level to the next. Theory, in general, fell into disrepute, and those who adhered to the evolutionary approach were scornfully dismissed as "armchair" anthropologists. As the new descriptionist and functionalist

Frederick Engels

Lewis H. Morgan

Edward B. Tylor

Robert Briffault

schools took over, what began as a unified panorama of social evolution became fragmented into a patchwork of field studies of contemporary primitive peoples. A world-historical outlook was reduced to a worm's eye view of bits and pieces of social customs and cultural traits.

The reasons given for this drastic turn away from the evolutionary approach come down to the objection that it is not possible, necessary, or desirable to reach back into ancient society, since all such attempts are no more than mere unscientific "speculations." This objection, usually accompanied by intricate and often incomprehensible technical details, makes it appear that the dispute is purely anthropological. But closer examination shows it to be also political.

The repudiation of the Morgan-Tylor school was provoked by their discomfitting discoveries that primitive society was a matriarchal, collectivist organization, featuring economic, social, and sexual equality. These findings exposed the myths that private property, the rule of the rich, and male supremacy had always existed, and that women were "by nature" the eternally inferior sex. As Engels pointed out, the downfall of women occurred when the primitive egalitarian society was overthrown by patriarchal class society a few thousand years ago. But this was not the end of the historical process. This degrading and oppressive society would in its turn be abolished and replaced by a higher form of egalitarian organization—socialism. Women, so long reduced to the second sex, would rise again to their rightful place in society.

These revolutionary conclusions were unacceptable to the new academic schools of anthropologists who took command of the science. As Haymes points out, Boas, Kroeber, Lowie, and their co-thinkers of the functionalist school changed the direction of anthropology, eliminating both the evolutionists and Engels. I explained twenty years ago in an article entitled "Anthropology Today," how, in their flight from the Marxists, the antievolutionists were also obliged to reject their own predecessors in anthropology. [This article, first published in 1957, appears in this volume under the title "Evolutionism and Antievolutionism."]

In that article, which was a critique of the Wenner-Gren funded *Anthropology Today,* I cited two eminent contributors to this collection. One was Grahame Clark who declared that the theories of Morgan, Tylor, and the other mid-Victorian anthropologists, as well as the Marxists, had "long ceased to be respectable" (p. 345). The other was Julian Steward who stated that the Marxist support of Morgan's work "has certainly not favored the acceptability of scientists of the Western nations of anything labeled 'evolution'" (p. 315). I summarized these political comments as follows:

> Here is the real reason for the antimaterialism and antievolutionism of contemporary anthropologists. The reactionary school has become predominant because it has accommodated itself to ruling-class prejudices and dogmas and assumed the obligation of stamping out the spread of revolutionary conclusions.

Thus the aversion to the evolutionists was politically motivated, although this was not usually openly expressed. Haymes himself does not refer to it. The politics were cloaked over by what were presented as anthropological disagreements. Some were the disputes that normally occur in the development of any new science, but others were subtle attempts to gut the evolutionary method employed by the pioneers.

Morgan's three broad stages of social history, from savagery through barbarism to civilization, were discarded on the grounds that he was a "unilinear" evolutionist. Similarly, the use of "survivals" in the reconstruction of prehistoric society was condemned as irrelevant and useless. Most students in the universities, intimidated by these anthropological formulas, were herded into the functionalist camp of anthropology.

Haymes recites these conventional arguments in his polemic. He states that up-to-date anthropologists are no longer concerned with "evolutionary theories, matriarchies, or 'survivals.'" This is unfortunately the case with most Western schools. But he also states that the "unilinear concept" survives only among the heirs of Engels, like myself. This is inaccurate on two counts. Morgan, Tylor, and Engels were not "unilinear" evolutionists, and other scholars besides the followers of Engels have kept alive the flame of evolutionism during the decades of functionalist retreat and reaction.

Let us examine these two major theoretical issues.

Survivals as a Record of History

Anyone who wished to find out about the evolution of human transportation would obtain this information for the most part from historical records. But no investigator would disdain the additional evidence to be gained from "survivals" of old and obsolete modes of transportation. Though the chariot of the Roman gladiators may be no more than a museum piece, the ox-drawn cart of early agricultural communities is still in limited use in the colonial world. The same is true of the horse and buggy which gave way first to the "horseless carriage" and then to the fully developed automobile. Thus some "survivals" of early transportation can be found alongside modern transportation. They are relics of more archaic modes of travel.

To a pure functionalist all these successive modes of transportation would only represent "varieties" of "particular" vehicles for transportation—which they indubitably are. But for an evolutionist they *also* represent a sequence of stages in the development of transportation. They show how humanity proceeded from foot to animal locomotion, to wheeled vehicles, to off-the-ground travel in airplanes and space ships. The evolutionist does not make an arbitrary dichotomy between the particular type of vehicle and the entire historical process: the particular is manifest evidence of the general in phenomena. Through survivals of its earlier forms, the investigator can trace the evolution of transportation from the beginning to the present point of development.

The same principle holds for other scientific investigations. In biology certain vestigial anatomical "survivals" in humans, such as the coccyx, testify to the primate origin of our species. In archaeology the "survivals" of fossil skulls, bones, and stone tools, dug up from the earth where they were preserved, enabled scientists to reconstruct the main milestones of human evolution from the earliest hominids to fully developed *Homo sapiens*. Anthropology, which, like archaeology, deals with history before the advent of written records, must also rely upon survivals for its reconstruction efforts.

These "social" survivals differ in form but not in essence from others. In place of fossil bones and tools they consist of ancient customs, practices, traditions, kinship relations, and clan structures. These survivals have been lost in civilized patriarchal nations but exist in primitive regions that retain some or all of their matriarchal features.

A graphic example is the matrilineal kinship system, which still survives in many primitive regions even though the rest of the world is overwhelmingly patriarchal. In most such instances today the paternal relationship between a woman's husband and her child is recognized, but in a weak, underdeveloped form. All the decisive kinship ties—descent, succession, and inheritance—pass through the maternal line in a matrilineal region. This matrilineal kinship represents a survival from the former matriarchal epoch. To toss out these vestiges of the past as irrelevant and useless can only serve to block off the way back to the matriarchy and an understanding of the place of women in that period.

Yet the modern descriptionist and functionalist schools refuse to recognize the role survivals can play in reconstructing social evolution. Robert H. Lowie is explicit on this score. In his book *The History of Ethnological Theory* he views survivals as "useless organs" and scoffs at the notion that these can serve any useful purpose either in anthropology or biology. "For in culture, as well as in biology, there may be alternative explanations for a 'useless organ,'" he writes. The "survival argument," he says, has hardly ever been demonstrated "outside of technology" (p. 26).

Even a child could figure out that the horse and buggy, a useless organ rarely seen on the streets of New York today, was once a useful vehicle that historically led up to the automobile. Thus, anyone wishing to be called a scientist would be hard put to rule out survivals in the study of technological evolution. But in the interpretation of anthropological data, it has been possible to do this and get away with it—for a time.

Lowie asserts that the matrilineal kinship system has no connection whatever with a prior matriarchal period. But he fails to explain how this peculiar matrikinship system came into existence at all if society has always been patriarchal. Nor does he indicate why it survives only in primitive regions possessing other matriarchal features, and never in civilized patriarchal nations. He simply dismisses Tylor and the other evolutionists with the sweeping statement that "the priority of 'mother right' as a general principle has thus been pretty generally abandoned" (ibid., pp. 43, 82).

To be sure, once the antievolutionists gained control over academic anthropology, studies of the matriarchy were abandoned, and the use of survivals as evidence of its former existence became unacceptable. Haymes, who sides with the functionalists, selects his own example from "linguistics" for conforming to the pattern laid down by his mentors. He objects to Morgan's use of "linguistic" survivals by which he "inferred" an early period of primitive society when paternity was unknown and sexual relations were "promiscuous." But the point is that not simply linguistic survivals but any kind of survivals are rejected if they lead back to the matriarchy. It is somewhat difficult to debate the pros and cons of anthropological data if the scientific method of the discipline has been gutted.

Ruling out survivals, however, left the descriptionists in a vulnerable position. Without evidence, survivalistic or otherwise, how could they prove their contention that patriarchal society, male dominance, and female inferiority have always existed? The more foresighted functionalists realized that some kind of accommodation had to be made to the evolutionary method for the sake of maintaining a scientific stance. This was achieved by making a restricted concession to evolutionism, coupled with the charge that any longer view of social history was a bad kind of evolutionism—"unilinear" evolutionism.

This charge was leveled at the nineteenth-century scholars who held to the "long view" of history. Julian Steward was one of the formulators of this thesis. "The inadequacy of unilinear evolution lies largely in the postulated priority of matriarchal patterns over the other kinship patterns," he wrote in his article "Evolution and Process" (in *Anthropology Today*. p. 316). He strenuously objected to acknowledging Morgan's first two sequences of social evolution, the periods of savagery and barbarism, which is not surprising since these were precisely the periods in which the matriarchy existed.

As against this invented "unilinear" approach, Steward put forward his notion of "multilinear" evolutionism. But this was nothing other than a patchwork of separate evolutions, restricted to the upper level of history, primarily the period of ancient civilization and its immediate antecedents. The functionalists regarded it as permissible to examine "particular" and "parallel" lines of evolution in different areas of the globe but not to bring them together as distinct aspects of a unified process going back to social origins on a world-historical scale.

Those who retained the evolutionist approach were sharply critical of these restrictions. Leslie White, in his article "Ethnological Theory," wrote that Steward "wants his evolution piecemeal . . . in restricted areas and in restricted segments . . . he anchors himself to the particular, to the local, or to the restricted" (*Philosophy for the Future,* p. 71). More recently another proponent of the evolutionary method, W. F. Wertheim, writes in his book *Evolution and Revolution* that Steward, the main representative of the multilinear thesis, "succeeded in restoring a new respectability to evolutionism" while evading a world-historical overview of human progress (p. 25). By maintaining the fragmented character of the evolutionary process in their schemes, the functionalists made a bow to the force of the facts without altering their objective of keeping the road back to the matriarchy blocked off.

Students in the universities have been taught to confine themselves to particulars and parallels in evolution but to abstain from drawing any larger historical generalizations. They were persuaded that a consistent evolutionary approach was invariably "unilinear" and therefore to be avoided. They were given little or no incentive to reinvestigate either the findings or method of the nineteenth-century scholars. Such studies would have detracted from their prospective careers as learned anthropologists. Under these conditions it was easy to also slip into blind acceptance of capitalism as the best of all possible social systems.

Haymes shows the effects of these influences. "I do not subscribe to the 'inevitability' of the Marx-Engels unilinear view of history," he says, and "the prospect of a 'unilinear leap' into collectivism is not desirable and certainly not inevitable." We might ask, how would Haymes feel about a "nonunilinear" leap from an oppressive capitalism that degrades women to an egalitarian socialist society? Would he consider this at least desirable, if not inevitable?

Actually, the allegation that the nineteenth-century pioneers were "unilinear" evolutionists is unfounded. It was introduced as a tricky device to do away with the significance of their discovery of the matriarchy.

The "Unilinear" Question

To a rigidly unilinear evolutionist all primitive regions throughout the world would have passed through the same stages of progress in a straight line of development—from a lower to a higher economy and culture. Such a view would not take into account the variations in cultures or the diffusion of cultures, that is, the borrowings of one culture from another. Nor would a unilinear evolutionist take notice of conjunctural setbacks or spurts forward of specific cultures in the overall evolutionary process.

This mechanical concept of evolution cannot be attributed to Morgan or Tylor, and even less is it the method used by Engels. Marxists, with their dialectical understanding of the historical process, are fully aware of its uneven development; the most pronounced irregularities of society persist to the present day. Side by side with advanced industrial nations, primitive peoples still survive in many pockets of the globe.

At the same time, and precisely through variations, diffusions, and combinations of all kinds, primitive regions can be profoundly affected by the more advanced nations. They can suffer setbacks, stagnation, or deformation under the pressures of more powerful, exploiting capitalist countries. But they also pass through combined developments in which they can telescope or skip over earlier stages of progress which other nations had already gone through.

For example, some primitive peoples, barely emerging from foot locomotion, have become directly acquainted with airplanes even though they have not independently passed through the intervening stages of the horse-drawn carriage, train, or automobile modes of locomotion which preceded air travel. Such uneven and combined forms of evolution have occurred throughout history, from savagery to civilization. Although Morgan and Tylor did not formulate the features of the process in this dialectical manner, they did recognize different ratios and tempos in historical progress and took account of variations and diffusions of cultures.

The functionalists, on the other hand, have drawn a barrier between diffusion and evolution, excluding the latter. Lowie makes the absurd statement that "diffusion plays havoc with any universal law of sequence" (*History of Ethnological Theory,* p. 60). It does nothing of the kind, and that is why Morgan and his co-thinkers were not guilty of such a dichotomy. They recognized "particular" cultures, "varieties" of cultures, and the "diffusions" of culture—but they placed these within the framework of a universal evolutionary point of view. They were not mechanical "unilinear" evolutionists.

As T. K. Penniman wrote in his book *A Hundred Years of Anthropology,* "Tylor was far from being a unilinear evolutionist with a simple and universal scheme of development." The influence of diffusion and culture contact was always present to his mind. At the same time, "he refused merely to cultivate a small field, and thereby to sacrifice the largeness of view that comes from attempting great and universal problems" (pp. 175-77).

More recently, in his book *The Rise of Anthropological Theory,* Marvin Harris, former chairman of the Columbia Department of Anthropology, challenged the claims of Lowie and the functionalists that Morgan's stages of evolution were "fixed sequences, every step of which had to be gone through by all cultures." In his sections on "The Myth of Unilinear Evolutionism" and "The Myth of the Denial of Diffusion," he shows how Morgan and Tylor have been "systematically misrepresented" with the label of "unilinear evolutionism" pinned on to them by Julian Steward. Although Harris chides the original evolutionists in his field for pushing their search too far—above all, in the direction of the priority of the matriclan system—he is far more critical of the "particularists," who "denied that a science of history was possible" (pp. 171-79).

Significantly, the taboo against applying the evolutionary method has not always been strictly observed even by some functionalists. This is the case with Malinowski, who probed around in the Trobriand Islands, where matriarchal survivals existed, and drew some important generalizations from his findings. This aroused the resentment of some of his colleagues. As Lowie wrote: "this scorner of history himself reconstructs the past," and added in biting condemnation, "he does so self-consciously, mumbling the purifying spell that he is 'discounting any undue antiquarian or historical bias'" (*History of Ethnological Theory,* p. 239). Yet Malinowski's utilization of the historical method—even on a partial basis—makes his researches immeasurably superior to those of the pure descriptionists.

Despite the near-monopoly exercised by the functionalists in academic anthropology, the evolutionary method could not be completely exiled. Indeed, a growing uneasiness has asserted itself over the anomalous position occupied by antievolutionary anthropology while sister sciences such as biology, paleontology, and archaeology continued on their evolutionary courses. The influence of Gordon Childe, the world-renowned archaeologist, helped to create a dissident current in the field of anthropology.

Prime credit for maintaining the evolutionary method in the United States must go to Leslie A. White, who kept the flame burning throughout the decades of deep reaction against it. Contrary to Haymes's statement that only the heirs of Engels have adhered to this method, there have been others besides the

Marxists—and they are growing in numbers. The White school has given rise to younger followers such as Marshall Sahlins and Elman R. Service, who edited *Evolution and Culture.* In his introduction to this book of essays, White wrote:

> All of the authors are younger anthropologists. . . . They were not reared in the atmosphere of antievolutionism; they accepted cultural evolutionism from the very start and have therefore been relatively free . . . to explore the implications of the theory of evolution as it applies to culture and to develop its many and fruitful possibilities. And they have done it exceedingly well [p. xi].

At that time, in 1959, White predicted that a reversion to evolutionism was inevitable. "The concept of evolution has proved itself to be too fundamental and fruitful to be ignored indefinitely by anything calling itself a science," he wrote. Since then the tide has been steadily turning in this direction. By 1974 the German scholar W. F. Wertheim, who surveys the period of deep reaction over the past few decades, opens his book with the summons: "We have to get back to evolution. When we threw out evolutionism, we threw out the baby with the bathwater" (*Evolution and Revolution,* p. 17).

These men must be given credit as scientists for maintaining their integrity and loyalty to the principle of evolution. However, it is also necessary to point out their shortcomings. Unlike the nineteenth-century evolutionists who disclosed the prior existence of the matriarchy, these twentieth-century defenders of their method are unable or unwilling, either through bias or indifference, to pursue this investigation any further. From the feminist standpoint, their works are deficient.

Neither Gordon Childe, the archaeologist, nor Leslie White, the anthropologist, despite their immense talents, were capable of discerning the role played by women as a distinct sex in the advancement of humankind in the first and longest epoch of evolution. Marvin Harris does not even list in his extensive bibliography Robert Briffault's *The Mothers,* which set forth the most fruitful theoretical contribution of the twentieth century—the matriarchal theory of social origins. Yet Harris's book deals with the rise of anthropological theory.

The deficiencies of these scholars throws light on what Haymes refers to as a "curious" new development—"feminist anthropology." The men have apparently exhausted their energies in defending evolutionism against the microanthropologists. Now the awakened and enlightened women have to carry the ball in the reconstruction of their past. The fresh breezes flowing through the stagnant enclosures of the functionalists will be fanned by the feminists, who will be sufficiently motivated to uncover and recover the full history of the female sex. My book *Woman's Evolution,* just off the press, is a contribution toward that goal.

Haymes's position on the feminist struggle corresponds to his functionalist position in anthropology. Although he insists that the Morgan-Engels approach has been declared old-fashioned and out-of-date, truth to tell, he is himself behind the times. Given the strides that women in the liberation movement are making, and the reopening of theoretical and historical questions concerning our past history, the advice Haymes is offering is rapidly becoming "dated."

His Advice to the Feminists

Haymes lays out five main areas of struggle for feminists: educational, artistic, legal, scientific, and ideological. The field of woman's history is conspicuous by its absence. But his plan has other defects.

Haymes believes that the "ideological" sphere of feminist intellectual activity is "easiest" to achieve. All it requires is for women to concur with Haymes's naive faith in American capitalism's capacity to produce "unparalleled wealth." However, even those feminists who have yet to read Engels are observing and experiencing the gross dislocations of capitalism in the grip of a severe crisis. They will hardly be inclined to place such blind faith in an economy that is in fact producing unparalleled profits for the big monopolists, while cutting down the opportunities of the middle classes and driving down the standards of the workers and the women.

Haymes's view of the "scientific" sphere is equally narrow and one-sided. He thinks that the unraveling of certain psychological myths, such as Freud's penis-envy thesis, removes the last bulwark against granting equal opportunity to women. However, no matter how many psychological myths are discarded, so long as capitalism profits from the exploitation of women they will remain the doubly oppressed sex, on the job and in home and family life. It is this *socioeconomic reality,* not psychological notions, that stands in the way of full liberation.

Having disposed of theory and science in a rather superficial manner, Haymes gets down to the practical activities for women. They should concentrate on the three remaining areas of work: educational, artistic, and legal. His suggestions for dramatizing the injustices of women who are denied equality—pressuring departments of education into nonsexist education and demanding legislation such as the ERA to fight against job discrimination—are acceptable and already being done. But Haymes goes wrong when he advises women to confine themselves to these activities, and he has illusions that these will suffice to give women all that they need and want.

Over the past six years the women in the liberation movement have made considerable gains in their practical activities. But Haymes is naive if he thinks that "a stroke of the pen" such as passage of the ERA—or any other legal or constitutional amendment—will "erase" discrimination against women. Such legislative and judicial strokes of the pen did not bring full freedom to Blacks, nor will they to women. Even now, women are obliged to defend the legal right to abortion—a right they won through unremitting struggle—against the powerful clerical forces campaigning to rob them of their victory.

Women know that they have to struggle to make practical gains here and now, and must also fight to retain them. Their accomplishments thus far testify to their capabilities in these areas. But this does not end the matter for growing numbers of feminists. They are determined to find out how women became the oppressed sex and whether, as they are constantly told, there is "no exit" out of an eternally inferior status. Women have a healthy suspicion that they will learn something different from their own history—once it is brought out of hiding. Feminists in league with feminist anthropologists intend to bring it to light.

This new trend of women reexamining their own history was featured in a November 2, 1974, *New York Times* story on the Berkshire Conference on the History of Women. More than two thousand teachers, graduate students, and other specialists "overran Radcliffe College and Harvard Yard for three days of intensive, sometime furious, discussion and debate." Under the headline: "The Woman in History Becomes Explosive Issue in the Present," the report stated: "Woman's multifarious part in history, long neglected or underestimated by traditional historians in the West—mostly male—is now being uncovered, publicized and promoted with extraordinary zeal. Indeed, the study of women in history is exploding in the academic skies like a supernova."

This is a harbinger of things to come. Uncovering their own hidden history is just as essential and urgent a part of woman's work as winning concessions and reforms in daily activities. Women will no longer accept the formula that theory and history are man's work while practical activities are chores for women. They will pursue their own intellectual course and will not cease their explorations until they find what they are looking for—the truth about woman's evolution.

The Challenge of the Matriarchy (1975)

Woman's Evolution, which deals with the hidden history of women, is a feminist book. But it is more than that; it marks a new theoretical turn in anthropology, which in recent years has witnessed a progressive deterioration in its methodology. Let us examine the reasons for this decline and what is required to put anthropology back on the right track.

Anthropology was founded by Morgan, Tylor, and other nineteenth-century evolutionists, who defined the new science as a study of prehistoric society and its origins. The two most important of the numerous discoveries they made were: Primitive society was a collectivist egalitarian system having none of the inequities of modern society, which is founded upon the patriarchal family, private property, and the state. It was likewise a matriarchal society in which women occupied positions of leadership in productive and social life and were held in high esteem.

These features stood in such sharp contrast to the conditions which prevail in patriarchal society that they soon gave rise to controversies which brought about a deep division in anthropological circles. After the turn of the century new trends of thought arose, led by Boas, Radcliffe-Brown, and others, who rejected the method and principal findings of the founding scholars—even while paying anniversary homage to them.

These schools abandoned a comprehensive evolutionary approach and substituted in its place empirical and descriptive field studies of contemporary primitive peoples surviving in various parts of the globe. They discarded Morgan's three stages of social evolution—from savagery through barbarism to civilization—without offering any pattern of progression of their

---

This article is based on a presentation Evelyn Reed made during a debate with Professor Walter Goldschmidt, past president of the American Anthropological Association, on May 24, 1975, at the University of California at Los Angeles. At his request, Professor Goldschmidt's rebuttal has not been published.

own. They say it is not possible or even necessary to go back to savagery, although this earliest historical period was by far the longest, comprising 99 percent of human life on earth. They foreshortened human history to the last ten thousand years or less of the million-year span of its evolution.

Some of those who formerly accepted this surgical operation are now complaining about the decadent state of anthropology. In a (London) *Times Literary Supplement* roundup covering the state of anthropology over the past fifty years, the British anthropologist Rodney Needham writes:

> Evolutionism was succeeded by diffusionism, which was supplanted by functionalism, and this in turn was superseded by structuralism; but after all these academic shifts and turns the state of secure understanding of man and his works has remained disappointingly static. . . . with increasing specialization and professionalism, social anthropology has actually become steadily duller and more trivial [July 6, 1973, p. 785].

As anthropology became more trivialized, further explorations into the matriarchal epoch and the hidden history of women virtually came to a halt. Students in the universities were taught that Morgan and the other founders of anthropology were "old-fashioned" and "out-of-date." In academic circles the matriarchy became a non-subject.

To justify this discrediting of the pioneers it is usually contended that there was "insufficient" documentation on the prior existence of the matriarchy, and, in any case, no one could ever draw any "universal" conclusions about a remote period that was forever closed off from view.

This contention is highly ironical since the opponents of the evolutionists have not hesitated to set forth some "universal" theories of their own. They say that the matriarchy never existed and the patriarchal family is eternal. They further say that women have always been the inferior sex, as they are today, because of their childbearing functions and other biological disabilities. Finally, they say that male supremacy has always existed because of the superior physical and mental abilities of the male sex.

How do they know these are "universal" phenomena if no one can ever penetrate into the facts about 99 percent of human life on earth? And where is the evidence to back up these assertions? Without such evidence their case is unproven.

The claim that there is insufficient documentation on the prior existence of the matriarchy is unfounded. The pioneer scholars brought forth a wealth of materials derived from different avenues of investigation. They assembled this data from literary sources as well as from actual observations and field studies on the matrilineal structure still surviving in many regions of the globe. They noticed that wherever matriliny was still in force patriarchal institutions were either nonexistent or only feebly developed. And they drew cogent conclusions from their studies of primitive customs, traditions, myths, and rituals which had survived from the former matriarchal epoch.

Unanswered Questions

If, as they argue, all this evidence is "insufficient," why have they so arbitrarily cut off further investigations of the subject so that more evidence could be obtained? And why, even on the basis of the evidence now available, do they refuse to answer certain questions on the priority of the matriarchy?

Let me pose three most important queries:

1. Opponents of the matriarchy do not deny the presence of the *matrilineal* kinship system, since it exists to the present day in many primitive regions. Where did this matrilineal structure come from if not from the ancient matriarchal epoch?
2. Why has the passage from matrilineal to patrilineal kinship *always* been in that direction, never the other way around?
3. Why is the ancient system of matrilineal kinship and descent found nowadays *only* in primitive regions and never in the advanced patriarchal nations, which have long lost and forgotten their matriarchal origins?

Those who fail to deal with these and related questions are merely asserting that the matriarchy never existed and the patriarchy is eternal. They have neither documentation nor adequate arguments to prove their contentions. Their method is prejudiced rather than scientific. It has produced more obscurity than clarity on the subject of human origins.

A leading American anthropologist, Marvin Harris of Columbia University, admitted this in a recent interview: "Sometimes I think that the primary function of establishment anthropology is to fog the truth" (*Psychology Today,* January 1975, p. 67).

Under these circumstances it is not surprising that anthropology has retrogressed in theory and method, even though additional valuable information on particular peoples has been accumulated. Once the academic anthropologists abandoned the evolutionary method and discarded the findings on the matriarchy, they not only cut off the major portion of our history but also any possibility of understanding the peculiar institutions and customs of matriarchal society.

Two of these are of primary importance: One is that curious institution called totemism; the other is the primitive kinship system. Early scholars devoted many years, some even lifetimes, to studying these subjects; their research shed considerable light on them, even though they were unable to answer all the questions their own studies had raised. But their successors, with their narrowly empirical method, have been even less able to fathom these fundamental phenomena of primitive life.

Turning Subjects into Non-subjects

This frustration had led some of them to decide that since they could not decipher these institutions they should be erased from the record. Totemism and kinship were thereupon dismissed as mere figments in the imaginations of earlier anthropologists. Thus, after having declared the matriarchy a non-subject, they went on to relegate its key institutions—totemism and kinship—to limbo.

Two examples should suffice to indicate how this vaporization of primitive institutions is taking place.

Rodney Needham, who is said to have published more original work on kinship than any other anthropologist, including the French professor Lévi-Strauss, declared in a recent book, "There is no such thing as kinship" (*Remarks and Inventions: Sceptical Essays about Kinship*).

Fred Eggan, reviewing this work for the December 13, 1974, (London) *Times Literary Supplement,* commented "this must be a difficult admission to make, since one does not waste twenty years on a non-subject without some emotional costs." Since the book deals with a non-subject, he characterizes it as a "non-book."

In the United States we witness an even more remarkable situation. A compilation commemorating the centennial of Lewis Morgan, the discoverer of the primitive kinship system, was published in 1972 by the Anthropological Society of Washington for its prestigious Smithsonian series. It is called *Kinship Studies in the Morgan Centennial Year*. It contains a paper by David M.

Schneider of the University of Chicago entitled, "What is Kinship All About?"

"The answer is simple and self-evident," he states. Kinship is an invention of Morgan's, and "*in the way in which Morgan and his followers have used it, does not correspond to any cultural category known to man*" (his emphasis, p. 50). Schneider thus agrees with Needham and writes that "there is no such thing as 'kinship.'" Since this is a complete turnabout from his previous position on kinship, he apologizes for the book he wrote in 1968 called *American Kinship* and says its title is a "misnomer." More to the point, he has converted his subject into a non-subject.

This is all the more surprising and dismaying because Schneider together with Kathleen Gough edited the compilation, *Matrilineal Kinship*, which appeared in 1961 on the centennial of Bachofen's work, *Das Mutterrecht*. Is this major work by Schneider and Gough now also a non-book on a non-subject?

Apparently Schneider is himself uneasy about this trend toward the evaporation of anthropological categories by renouncing anthropologists. But his attempt to justify or even explain the repudiation borders on the incredible. He writes:

> For a while anthropologists used to write papers about Totemism. . . Goldenweiser and others then demolished that notion and showed that totemism simply did not exist. . . . It became, then, a non-subject. In due course Lévi-Strauss wrote a book about that non-subject, in which he first explained that it was a non-subject and therefore could not be the subject of the book. . . . The "matrilineal complex" suffered the same fate in the hands of Lowie.
>
> In my view, "kinship" is like totemism, matriarchy, and the "matrilineal complex." It is a non-subject. . . . If you like to think that I have devoted a good part of my intellectual life to the industrious study of a non-subject, you are more than welcome to do so. If you think that I have now talked myself out of a subject for study you are quite right, too ["What is Kinship All About?" in *Kinship Studies in the Morgan Centennial Year*, p. 50].

Such is the sad state of anthropology today in the hands and heads of confused and despairing men. They frankly admit that, one by one, the most significant subjects have been degraded to non-subjects and their books reduced to non-books. Are we now to look forward to the final admission that anthropology itself is bankrupt and a non-science?

Indeed, Robin Fox, a co-thinker of Needham and Schneider and author of the book *Kinship and Marriage* has now come to question what anthropology is all about and how he fits into it. He opens his later book *Encounter with Anthropology,* with the following confession:

> This is a book about anthropology by a puzzled anthropologist who does not know quite how he fits into his discipline any more. And judging by the book reviews of some of his colleagues, the discipline has *its* doubts about the relationship. Something is wrong somewhere [p. 9].

This is the end result of the wrong course taken more than half a century ago by the antievolutionists. It has brought about the stagnation and demoralization of a once-vigorous branch of social science. Some anthropologists, like Evans-Pritchard, are alarmed by this confirmation of a warning once issued by Maitland that "by and by anthropology will have the choice of being history or being nothing" (*Kinship Studies in the Morgan Centennial Year,* p. 14).

What is needed to rescue anthropology from its blind alley? It must return, although on a higher level, to the evolutionist and materialist approach of the pioneer scholars.

That is precisely what I have tried to do in my book *Woman's Evolution,* which begins with the basic premise of the priority of the maternal clan system or matriarchy. Upon this foundation I have been able to develop new theories that explain the meaning and purpose of certain enigmatic institutions of savage society, including the ones so cavalierly disqualified as non-subjects—totemism and kinship.

Totemism and Taboo

My own theory on totemism came about, somewhat accidentally, through a closer examination of taboo, which is indissolubly connected with totemism. I could not accept the standard reason given for the primitive taboo—that it was directed against incest. Primitive peoples were ignorant of the most elementary biological facts of life, including how babies are conceived and the inevitability of death. How, then, could they have understood the concept of incest, which presupposes a very high degree of scientific knowledge? Genetics, for example, is only as old as this century.

Moreover, the taboo was a double taboo, applying to food as well as to sex. In fact, the clause applying to food was the more elaborate and stringent prohibition. Most investigators were aware of this twofold character of the taboo, but because the food prohibition seemed inexplicable, they focused their attention on the sex clause, and assumed that it was directed against incest. But the very fact that it was as much a food as a sex taboo ruled out this assumption.

But this raised the question: why so stringent a food taboo? Since the taboo is regarded as the oldest prohibition in human history, going back to the primeval epoch, the thought entered my mind that it must have been directed against cannibalism. There was a logic to this surmise which was confirmed by my subsequent researches. Apes in nature are vegetarians; our branch of the primates became meat-eaters only after they became hominids. How could they know, at a time when they were still part ape, that all hominids belonged to a species distinct from all other animals?

In other words, the earliest hunters had to learn what flesh they could *not* eat at the very time they were learning how to hunt, kill, and eat flesh foods. This dilemma, stemming from biological ignorance, could only be solved through social and cultural means. To my mind, this explains how the first social regulation in human history—totemism and taboo—developed. It was first and foremost a prohibition against cannibalism, and it began as a protection of the totem-kin.

Totem-kinship marked the dividing line between human flesh that could not be killed or eaten and animal flesh that could. Those who were born of the same horde of mothers and who lived and worked together in the same community were the totem-kin; that is, the human beings, the "people." Outsiders and strangers were non-kin and therefore nonhuman; they were "animals" which could be killed and eaten. Thus, while totem-kinship began on a small and limited scale, it furnished protection for the in-group, or kin-group—the primal horde.

Subsequently this protection against cannibalism became broader in scope. This was accomplished through the interchange system, usually called "gift-giving," by which different groups began exchanging food and other things with one another. These acts converted them from strangers and enemies (or "animals" in the most primitive concept) into new kinds of kinsmen and friends. These linkages created a network of affiliated clans which ultimately became the large tribe. On this higher cultural level cannibalism dwindled to an occasional ritual until it vanished altogether.

Not an Incest Taboo

The other clause of the taboo was simply a sex taboo, having nothing whatever to do with incest. As many scholars have pointed out, male sexuality in the animal world—where males fight one another for access to females—is a violent force. Such individualism and competitiveness had to be suppressed since human survival depended upon the closest cooperation of all the members of the group. Thus, it became imperative to overcome animal sexuality and to convert fighting males into the human brotherhood.

This goal was also achieved through totemism and taboo. All males in the totem-kin group were forbidden access to any females of that group. All the older women were classified as the mothers (or older sisters); the women of a man's own generation were his sisters, and the female children were his younger sisters. In this way the antisocial characteristics of animal sexuality were suppressed, and the foundation for the tribal brotherhood was laid. The clan system of social organization arose as a *nonsexual*, economic and social association of mothers, sisters, and brothers.

The two clauses of the taboo have an interlocking relationship. Food and sex represent the most imperative hungers in human and animal life; they are the twin driving forces behind the survival of the species. The hunger for food must be satisfied if the organism is to maintain itself; the hunger for sex must be satisfied if the species is to reproduce itself. The double taboo on food and sex therefore represents the earliest social controls over these imperative needs. And without these controls, human organization could not have gained its start.

Far from being a figment of the imagination of early anthropologists or a non-subject, totemism is in fact one of the most important subjects to be investigated in reconstructing our most ancient history. Totemism and taboo represent the means by which humankind elevated itself out of animality. Totemism is all-inclusive; it not only represents the totem-kin and the totem protectorate against cannibalism but also all the social regulations that were required for humanizing the species. Through totemism and taboo, humanity survived and thrived until it could reach a higher stage of social and cultural life.

Those who have turned away from the matriarchy, however, fail to understand totemism because it was the female sex that instituted it. Contrary to current myths about their status, women have not always been the inferior sex as they are today. In the beginning the females were the advantaged sex; they were the mothers, responsible for the survival of the species. Unlike males, who suffered from the biological handicap of incessant striving for dominance over other males, females could band together for the protection of themselves and their offspring. This nurturing, cooperative trait enabled the females to make the great advance from the maternal brood in the animal world to the maternal clan system in the human world.

Then, through the institution of totemism and taboo, the females were able to correct the biological deficiencies of the males. They began by socializing the two basic hungers. They expelled all internal hunting—whether for food or mates—from the group composed of totem-kin mothers, sisters, and brothers. By this means, both cannibalism and fights for dominance were overcome, and males were brought together as the clan brothers. This cooperative association of men—the fratriarchy, as I call it—has no counterpart in the animal world. It represents the crowning achievement of the totemic system, which was instituted by the women.

The Primitive Kinship System

Now let us examine the primitive kinship system, which skeptical anthropologists want to reduce to a non-subject.

The kinship system in its mature form grew up out of the totem-kinship system. The difference in development lies in the fact that the totem-kin included certain animals along with humans, whereas the mature system was restricted to humans alone. Lewis Morgan called this system the "classificatory" kinship system to distinguish it from the kinship system which exists today.

Classificatory kinship was a system of *social* kinship embracing all the members of the community, whereas our family system is restricted to the genetic members of the same family circle. In other words, all the members of the clans and affiliated clans were *social* mothers, sisters, and brothers, their biological relationships being unknown or irrelevant.

Like its predecessor the totem-kinship system, the classificatory system was also matrilineal; that is, kinship and descent were traced through the maternal line. However, the male line of kinship and descent in the matriarchal period was traced through the "fraternal" line, i.e., the mothers' brothers. This represents the "missing link" in fully understanding the matrilineal kinship system, which was also fratrilineal.

In the course of time, patrilineal kinship also became recognized when the man who married the mother became the father of her child. About eight thousand years ago, what Morgan called the "pairing family" (the father living under the same roof with the mother and her child) came into existence. Gradually the father and patrilineal kinship crowded out the mother's brother and fratrilineal kinship. However, it was not until the fully developed patriarchal family displaced the pairing family that the classificatory system of kinship was overturned and replaced by the family system of kinship.

It is not possible to understand the primitive kinship system without taking into account these evolutionary sequences. The worst errors have been made by those who declare that the patriarchal family has always existed; they write articles and books about "patrilineal bands" and "patrilineal clans" as though these were identical with the patriarchal family. In fact, the mere recognition of patrilineal kinship did not alter the basic structure of the clan as essentially matrilineal and fratrilineal.

Patrilineal kinship in the period of the matriarchy was no more than a paternal relationship between two matrilineal clans in which the mother-brother relationship remained preeminent and decisive. In other words, every so-called patrilineal clan was also and more fundamentally a matrilineal clan with the mothers' brothers occupying a more important and permanent status than the newly emerging husbands and fathers.

To understand this more clearly, a correction has to be made in describing the matrilineal kinship system. Usually this system is taken to mean that descent was originally traced through the mother-line alone. However, while descent in general was traced through the maternal line, precisely because the clan was a mother-brother clan—*male descent was traced through the mother's brother-line.*

This may come as a shock when we consider that under the taboo a clan brother could not marry a clan sister and therefore could not be the biological father of his sister's son. However, when we remember that primitive people were ignorant of biological paternity and that primitive kinship was exclusively *social* kinship, we can see that the mothers' brothers were just as capable of performing various functions for their sisters' children as the husbands and fathers who later took their place.

As the anthropological record shows, the mothers' brothers were the guardians and tutors of their sisters' sons—and the male line of descent, succession, and inheritance accordingly passed from maternal uncle to nephew. This line of descent prevailed throughout the entire epoch of the matrilineal clan system, even after patrilineal kinship was recognized. However, while it was possible to assimilate patrilineal kinship into the mother-brother clan without altering its basic structure, the same was not true of patrilineal *descent*. Changing the line of male descent from mother's brother to father shattered the fratriarchy—and that, in turn, brought down the matriarchy. Both were replaced by the patriarchy.

It is not possible here to go into all the factors that led to the drastic overturn of the mother-brother system by the father-system. The economic factors that led to the rise of private property, class divisions, and the downfall of women have already been spelled out in detail by Engels and others, and I have added some further data in *Woman's Evolution*. What concerns us here are the contradictions which developed in the relations between mothers' brothers and fathers, preventing an easy transition from male descent through the mother's brother-line to the father-line.

The "Divided Family"

In my book I explain that the matrifamily (my term for Morgan's "pairing family") was the last stage in the evolution of the matriclan system. Because it recognized the father and patrilineal kinship, it was a "divided family.' It was torn between two functional fathers—the mother's brother and her husband. However, the mother's brother held the fixed, permanent, and traditional ties to his sister's son, while the father had only ephemeral kinship ties to his wife's child. A son was kin to his father only as long as his mother's marriage lasted—and these marriages were easily and often broken.

Thus, while patrilineal *kinship*—of this weak and subordinate type—could be accommodated within the framework of the matrifamily, patrilineal *descent* could not. A child's relationship to his mother's brother was a "blood" relationship in the primitive sense of that term. This meant that in all critical situations, above all blood revenge, the son stood on the side of his mother's brother, not that of his father. Along with these "blood" obligations, the son's line of descent, succession, and inheritance were passed down to him from his maternal uncle, not his father.

This perpetuated the division inherent in the "divided family" and was a serious obstacle in the path of the full development of a unified one-father family.

To us, in hindsight, it may seem like the easiest, most logical concession in the world for the mother's brother to resign his place in his sister's family, give up his matrilineal fathership of his sister's son, and move on to become the patrilineal father of his wife's son. But that is not the way it worked out at that historical turning point in the transition from the divided family to the one-father family.

The chains of tradition and custom bound the participants of that period. They did not know how or why they had inherited their one-sided matrilineal kinship system, nor did they know how to liberate themselves from it once it became outworn and obsolete.

The result was a protracted and bloody struggle between the contending categories of men—the matrilineal and the patrilineal fathers. I describe this transition in *Woman's Evolution*, which details the extremely painful process by which the divided family finally was replaced by the patriarchal one-father family. In the same process the family system of kinship replaced the former classificatory, or social, system of kinship.

My book ends with a fresh analysis of three great Greek tragedies about the myth-histories of Medea, Oedipus, and Orestes. All reflect in dramatic terms the terrible cost paid to achieve the victory of the patriarchal family. In other words, far from being eternal, the father-family came into existence only a few thousand years ago. And the primitive kinship system, far from being a non-subject, is precisely the subject that explains the birth of the father-family.

This is only a brief review of my theories on the key questions of matriarchy, totemism, and kinship. Using the evolutionary method, I have tried to rescue these subjects from oblivion. But there are many other topics covered in *Woman's Evolution*. I present new insights into what are obscure areas of investigation. These include such phenomena as exogamy-endogamy, parallel kinship and cross-cousin marriage, the rites of passage with reference to initiation and couvade, the origin of the blood-revenge system, and the interchange system usually called "gift-giving."

Hopefully, these explorations will send fresh breezes through the science of anthropology and help remove the pessimism that afflicts its outlook today. Under the impact of the women's liberation movement, many areas of our lives and history are taking on a new look. Not least among these is the earliest history of women, which can be recovered only through the avenue of anthropology.

The Misconceptions of Claude Lévi-Strauss
On "The Elementary Structures of Kinship" (1977)

In a well-known folktale by Hans Christian Andersen called "The Emperor's New Clothes," the populace assembled to gaze upon and admire the new royal raiment of their ruler. Suddenly the humbug performance was exposed when a little girl cried out, "But he has no clothes on at all—the Emperor is naked!"

After his major work, *The Elementary Structures of Kinship*, was published in 1949, Claude Lévi-Strauss rapidly ascended the ladder of eminence to become the unofficially crowned head of anthropology. But by the 1970s mounting disillusionment and criticism make his place on the peak somewhat precarious.

This was brought out in a March 20, 1977, (London) *Times* article entitled, "Lévi-Strauss and the Marriage Market." It stated that in a question put to a number of pundits by the (London) *Times Literary Supplement* "to name the most overrated of current intellectual or literary eminencies, the name of Claude Lévi-Strauss duly appeared." This drastic change, from the "intense adulation and influence which Lévi-Strauss enjoyed in the nineteen-sixties," was due to the fact that the anthropological evidence on which he built his theories "is either scant or will not stand up to rigorous verification." The article adds that his very "brilliance as a prose-stylist" both masks and reveals this "emptiness of substance."

This negative view of Lévi-Strauss did not, however, extend to his antiwoman stance. The article sums up Lévi-Strauss's views on this matter: "in different societies and kinship systems, marriage is a device whereby women literally 'circulate' between familial and tribal groups. They are 'exchanged' in just the way in which goods are exchanged in trade." This theme of women as "currency," it adds, has been "powerfully" extended into other disciplines such as sociology, economics, and linguistics, where it has taken hold and "proved fertile."

So, whatever discontents the pundits have with Lévi-Strauss, they still share a common bond—upholding male supremacy. Women, it seems, are objects, to be dominated and manipulated by men. Lévi-Strauss is the leading fabricator of the myth that women have always been bandied about by men like pieces of property.

Let us examine Lévi-Strauss's scheme of the elementary structures of kinship—first his method, and then his assertions about women, which he bases on misinterpreted data.

His Antihistorical Method

It is curious that Lévi-Strauss dedicates his *magnum opus* to Lewis Morgan since the two men stand on opposite sides of the Great Divide separating the nineteenth-century evolutionists from their twentieth-century opponents. Morgan's method was substantially materialist and historical, and by applying it he discovered the maternal clan organization which preceded our patriarchal family system. Lévi-Strauss, on the other hand, is idealistic in his approach and belongs to the unhistorical schools which long ago abandoned the method and findings of Morgan and his co-thinkers. Lévi-Strauss nowhere mentions the maternal clan system, or "matriarchy" as it is also called, in this or his other books. Why, then, the homage to Morgan?

Perhaps it is because Lévi-Strauss fancies himself a Marxist thinker; indeed, in Europe this is almost obligatory for a man aspiring to a place in vanguard intellectual circles. This obliges him to tip his hat to Morgan, whose work was acclaimed and used by Marx and Engels. Lévi-Strauss, who would like to be all things to all men, prefers to repudiate the historical method indirectly rather than openly. His chapter "The Archaic Illusion" is a prime example of this artful dodging.

Lewis Morgan arranged the sequence of social evolution in three main stages: from savagery through barbarism to civilization. The earliest is sometimes referred to as the infancy of humankind, analogous to the child who must pass through that stage before becoming an adult. Lévi-Strauss does not openly challenge Morgan's sequential stages; rather he opines that a historical study is not sufficient for his purposes and that "certain fundamental structures of the human mind" must also be taken into consideration. It then turns out that these "mental structures" are essentially the same—good for all history and all societies (*Elementary Structures of Kinship,* p. 84).

Is there an archaic period of history with a primitive social structure that differs radically from ours, as Morgan pointed out? Lévi-Strauss adroitly shifts the axis of the discussion from social and historical structures to unchanging mental structures. He writes, "it is indeed tempting to see primitive society as approximating more or less metaphorically to man's infant state" (ibid., p. 88). Then he strays off, comparing the minds of contemporary primitive children and adults with civilized children and adults—a topic irrelevant to the subject under discussion.

To Lévi-Strauss it is the invariant mental structures that give rise to social structures, not the other way around. He writes that "each type of social organization represents a choice, which the group imposes and perpetuates" upon every newborn child (ibid., p. 93). Thus, primitive people chose their social structure just as civilized people chose theirs. "Undoubtedly, there are some differences between primitive and civilized thought," he writes, "but these are due solely to the fact that thought is always 'situational'" (ibid., p. 91).

In the end, after working our way through numerous psychological and other irrelevancies, we learn that there is no "archaic 'stage' in the intellectual development of the individual or species" (ibid., p. 97). The only conclusion to be drawn from this is that there is no archaic stage in social history—that such a stage is only an "illusion." Hence, his chapter title, "The Archaic Illusion."

These static, mentalistic views of Lévi-Strauss have nothing in common with the thought of Morgan and other evolutionists. They saw society developing out of the material conditions of life and changing with the advances made in these material conditions. Historically, human thought, like society itself, has grown up from the infantile to a higher, or adult level. For example, primitive people resorted to magic to explain phenomena, whereas we explain them through science.

Despite his apparent homage to Morgan, Lévi-Strauss is constrained to disavow Morgan's method at regular intervals in his book. These disclaimers appear whenever he is discussing the institutions of primitive society. Thus, on the dual organization, he writes, "we have been careful to eliminate all historical speculation, all research into origins, and all attempts to reconstruct a hypothetical order in which institutions succeeded one another" (ibid., p. 142).

On primitive kinship and marriage, he writes, "In the course of this work we have steadfastly avoided historical reconstructions" (ibid., p. 459). Elsewhere on the same point he says, "there is no need for us to reconstruct some archaic state in which Indo-European society practised cross-cousin marriage, or even recognized a division into exogamous moieties" (ibid., p. 472). He does not explain why the archaic institutions of dual organization, exogamous moiety, and cross-cousin marriage have survived up to recent times in primitive regions, even though they have become extinct in the civilized nations. Rather, he says, with respect to these matrilineal survivals, "it is possible, and even probable, that sociologists who defend the theory that all human societies have passed from a matrilineal stage to a patrilineal stage have been victims of an optical illusion" (ibid., p. 409).

Lévi-Strauss's dismissal of the historical approach to prehistoric society was set down in 1949 when the antievolutionist crusade in anthropology was at its height. But thirteen years later, having arrived at the top, he found it necessary to defend his antihistorical approach and even modify it.

In *The Savage Mind* Lévi-Strauss discusses his quarrel with Jean-Paul Sartre on the question of the dialectical and historical method, since "in both our cases Marx is the point of departure of our thought" (p. 246). Sartre, in his fashion, defends the historical approach, whereas Lévi-Strauss equivocates. Lévi-Strauss complains that Sartre is too engrossed in history and has an "almost mystical conception" of it. By contrast, he writes, "the anthropologist respects history, but he does not accord it a special value" (ibid., p. 256).

The "special value" apparently comes from attempting to reconstruct the line of development in the whole history of humankind. According to Lévi-Strauss this is an impossible enterprise not worth pursuing. The reason he gives for this aversion to synthesis is that of pure relativism. He writes, "Every corner of space conceals a multitude of individuals each of whom totalizes the trend of history in a manner which cannot be compared to the others" (ibid., p. 257). Therefore, a "truly total history" would confront the historian with "chaos."

To Lévi-Strauss only segments, or pieces of history count. He writes:

> In so far as history aspires to meaning, it is doomed to select regions, periods, groups of men and individuals in these groups and to make them stand out, as discontinuous figures, against a continuity barely good enough to be used as a backdrop. A truly total history would cancel itself out—its product would be nought. . . . History is therefore never history, but history-for. It is partial in the sense of being biased even when it claims not to be, for it inevitably remains partial—that is, incomplete—and this is itself a form of partiality [ibid., pp. 257-58].

If this sounds somewhat unintelligible, the main point is clear; Lévi-Strauss is partial to partial histories and opposed to a total overview of human society from the hominid stage to the beginning of civilization.

As an additional argument against seeking historical continuity, he offers the following: "There is no history without dates" (ibid., p. 258). Since we are acquainted with only a fragment of dated history, it is a

> discontinuous set composed of domains of history, each of which is defined by a characteristic frequence and by a differential coding . . . It is no more possible to pass between the dates which compose the different domains than it is to do so between natural and irrational numbers. . . . It is thus not only fallacious but contradictory to conceive of the historical process as a continuous development, beginning with prehistory coded in tens or hundreds of millennia . . . [ibid., pp. 259-60].

Levi-Strauss thus opens to question most archaeological findings of fossil bones and tools, which so far have rarely been dated precisely. This will hardly deter archaeologists and paleontologists from developing improved methods and skills for more precise datings.

Lévi-Strauss is dutifully abiding by the restrictions which prevail today in anthropology with its patchwork of isolated field studies and piecemeal historicity. This procedure is contrary to that of Morgan, Tylor, and the other evolutionists who made the first strides toward reconstructing the whole march of humankind.

But they had the bad judgment to disclose their discoveries about the matriarchy and the high position of women in it. They made these discoveries because they were not averse to applying a "total" historical method, which is not the case with the current schools. Lévi-Straus himself facetiously remarks, "As we say of certain careers, history may lead to anything, provided you get out of it" (ibid., p. 262).

Levi-Strauss's *magnum opus* is designed to tell us about the "elementary structures of kinship." But without the historical approach he is not equipped to do this job, precisely because these structures belong to the "elementary" stage of history—the archaic or prehistoric period. This can be seen when we examine them from the evolutionary standpoint, in contrast to Lévi-Strauss's static "structural" approach.

Maternal Clan Kinship Versus Father-Family Kinship

The elementary structures of kinship are not to be found in the father-family, the unit of patriarchal society, but in the maternal clan, the unit of primitive matriarchal society. The maternal clan was not composed of family relatives but of sisters and brothers in the social or collective sense. This is called the "classificatory" system of kinship.

Under that system all the members of the clan were classified by sex and by age or generation. Thus the older women were classified by sex as the "mothers," and by generation as the "older sisters" to the "younger sisters" in the age-group below them. Similarly on the male side, the older men were classified by sex as the "mothers' brothers" to the "younger brothers" in the generation below them.

Originally there were no mates residing in the maternal clan unit. These mates lived in their own maternal clan units where their social classification was also that of sisters and brothers. Before the paternal relationship was recognized, the mothers' brothers (or maternal uncles) performed the functions of fatherhood to their sisters' children. Kinship and descent were reckoned exclusively through the maternal line, with a subsidiary male line passing from mothers' brothers to sisters' sons.

It is this maternal clan kinship system, or the "classificatory system," which represents the "elementary structures of kinship." As the early scholars put it, "fathers were unknown," and only the kin on the maternal side were recognized. Thus in the older generation there were the mothers and mothers' brothers, and in the generations below them the sisters and brothers.

However, this is only the clan's internal structure. It also had external affiliations and alliances with other clans after it had evolved out of the primal-horde stage of its existence. Through these external relations with other clans, there gradually arose a whole network of clans and phratries which comprised the large tribe.

These external relations evolved in two directions, both adding further kinship categories to the classificatory system. One is the category of "parallel cousins" and the other of "cross-cousins." These differ with respect to the taboo that prohibited sex relations, or "marriage" as it is usually called, between the clan sisters and brothers. Under the parallel affiliations, which came first in history, these parallel cousins came under the same taboo. A man could no more marry his female parallel cousin than he could his own clan sisters.

It was quite otherwise with the cross-cousin alliance, which emerged at a later point in history. Cross-cousins were permissible mates in a regulated exchange-mating system between the clans and phratries involved. As this is often expressed, the sisters and brothers of one side found their mates among the brothers and sisters of the other side, and vice versa.

These external affiliations and alliances represent a higher stage in the evolution of the elementary structures of kinship. They were effected through what is often called the primitive "gift-giving" system, by which various necessities of life were given and received as evidence of the pacts made between the clans and phratries.

These elementary structures of kinship are also known by institutional terms. Thus the cross-cousin mating alliance is known as the "dual organization." This indicates the juxtaposition of the two sides of sisters and brothers vis-à-vis their cross-cousins on the opposite side. The dual organization also represents "exogamy-endogamy." Since the sisters and brothers of each side are "exogamous," that is, cannot mate with each other, they are "endogamous" with respect to their cross-cousins on the other side.

In summary, then, the elementary structures of kinship come down to the classificatory system, exogamy-endogamy, cross-cousin mating, the dual organization, and the gift-giving system that furnished the cement binding these building blocks together into the network of clans and phratries comprising the tribe. These structures and their evolutionary sequence are spelled out in *Woman's Evolution*, which also shows that *all of them belong to the maternal clan epoch, or matriarchy.*

Lévi-Strauss has a quite different point of view. Although it is to his credit that he reexamines every one of these institutions, which have been largely neglected for several generations, he does not have a single reference in his book to the maternal clan. With his antipathy toward the historical approach, he sees these institutions through the prism of modern patriarchal institutions and relations. He does not view matrilineal kinship and descent as a survival from the matriarchal epoch. Rather, he sees it as something haphazardly coexisting with patrilineal kinship and descent and always subordinate to the latter. This flows from his fundamental position that the father-family is eternal and there never was a matriarchy.

Having disposed of the maternal clan, Lévi-Strauss proceeds to contruct his own curious edifice of the elementary structures of kinship. He begins with the father-family, pair-marriage and "affines" (relatives by marriage). This confronted him with so many difficulties that, in a way, he is still fussing with the starting point of his construction.

The initial definition of what he means by "elementary structures" appears in the first paragraphs of his preface to the first edition. It is so vague and jumbled that eighteen years later, in the preface to the second edition (1967), he felt obliged to respond to his critics on this and other obfuscations. However, he makes things no clearer in the latter one.

More recently he ventured another attempt. In an article entitled "Reflexions Sur L'Atome de Parenté" in the July-September 1973 *L'Homme*, the anthropological journal published in Paris, Lévi-Strauss discusses what he calls the "atom of kinship." He first gives us the version he constructed in 1952 at the Bloomington, Indiana, Conference of Anthropologists and Linguists: namely, that a truly elementary structure of kinship, an "atom" of kinship, "consists of a husband, a wife, a child, and a representative of the group from which the first has received the second."

Who is this unnamed phantom "representative" who gives the woman to the husband? Presumably it is the mother's brother who is said to hand over his sister or sister's daughter in marriage. In that case, a male figure belonging to the maternal clan, the mother's brother, is added to the patriarchal atom of father-mother-child.

In the later 1973 version, however, the male component of the "atom" is enlarged. It is now the "quadrangular system" of relations between brother and sister, husband and wife, father and son, maternal uncle and nephew. Conspicuous by her absence is the mother in this male-dominated atom of brother-husband-father-son-uncle-nephew—with a wife and sister thrown in. This, in his opinion, is the "simplest structure" that anyone can conceive. It is indeed simple—in the sense of being absurd. He combines maternal clan categories of relationships, torn out of their historical context, with the relatives of the modern patriarchal family circle.

Lévi-Strauss does not view the mothers' brothers, or maternal uncles, as historical figures from an archaic period when they were in charge of their sisters' sons, before the father-family came into existence. In *Structural Anthropology* he scoffs at E. Sydney Hartland whose signal achievement was to bring forward the key role played by the mothers' brothers, a role second only to that of the mothers themselves. He takes the official line of anthropology today and rejects Hartland along with the other evolutionists, because to them "the importance of the mother's brother was interpreted as a survival of matrilineal descent." To Lévi-Strauss, "this interpretation was based purely on speculation, and indeed it was highly improbable in the light of European examples" (*Structural Anthropology*, p. 37).

As against Hartland, Lévi-Strauss applauds Lowie for his "merciless criticism of evolutionist metaphysics." According to Lowie the prominent position of the mother's brother "cannot be explained as either a consequence or a survival of matrilineal kinship" (ibid., pp. 37-38). So, taking his cue from Lowie, Lévi-Strauss converts these clansmen into individual brothers, uncles, and nephews, adding them to the father-son atom of the patriarchal family.

Along with his practice of reducing clan kinship categories to family relatives, Lévi-Strauss is not averse to misinterpreting his sources, particularly the writings of the pioneer scholars. One of his colleagues, E. R. Leach, writes, "I am bound to state that Lévi-Strauss has often seriously misunderstood his sources and that in important particulars this misunderstanding is due to quite inexcusable carelessness" (*Rethinking Anthropology*, p. 77).

Whether through carelessness or calculation, Lévi-Strauss not only misinterprets his sources but gives misleading translations of primitive terms. One pertinent example is to be found in his book *Totemism*, where he writes that the Ojibwa term *ototeman* comes from the word *totem* in the Algonquin language. The expression, he says, "means roughly, 'he is a relative of mine'" (*Totemism*, p. 18).

This translation is too general; it can be applied equally as well to a modern family relationship as to an ancient clan relationship. In fact, *ototeman* is a classificatory term that applies only within the maternal clan kinship system. It is translated by scrupulous authorities as "his brother-sister kin" or "his own clan," reckoning kinship through the female line.

Although this is a small instance compared to the big misinterpretations, they are all directed to one end: to conceal the historical priority of the maternal clan system and, along with it, the leading part played by women. This is apparent in Lévi-Strauss's one-sided and distorted interpretation of cross-cousin marriage as no more than "sister exchange."

Cross-Cousin Marriage and "Brother Exchange"

The institution called "cross-cousin marriage" has been a puzzling one from the time it was discovered. Under the classificatory system of kinship there are two categories of "cousins": the one called "parallel cousins," the other "cross-cousins." Neither category bears any resemblance whatever to our own system today in which cousins are genetically related to a specific family by "degrees," i.e., first, second, and more remote cousins.

Under the rules of the classificatory system, "parallel cousins" are prohibited from mating with each other, just as brothers and sisters are. Brothers and sisters and parallel cousins all find their mates among their cross-cousins on the opposite side of the dual organization. This is the puzzler. As some anthropologists put it, "Why cross-cousin marriage?" Others have wondered, "Why the dual organization?" But this is only another way of making the same inquiry. Even the pioneer scholars were unable to provide a full answer to this riddle.

Although Lévi-Strauss is just as puzzled, he purports to give an answer by examining this maternal clan institution through the prism of modern family ties. To him both cross-cousins and parallel cousins are essentially no different from the "first" cousins of today. He writes in *Elementary Structures of Kinship* that "The moiety system divides all first cousins into cross and parallel cousins" (p. 166).

But he refuses to recognize that the "moiety system," or the dual organization as it is more often called, belongs to the matriarchal epoch. How can a modern family relationship of "first cousins" be divided up into the archaic relationships of parallel cousins and cross-cousins—the terms for which do not even appear in our dictionaries?

Far from giving an adequate answer, Lévi-Strauss confuses the record still more with a fraudulent version of the "postulates" of the scholars of the "latter half of the nineteenth and early twentieth centures." He writes: "These postulates may be summarized as follows. A human institution has only two possible origins, either historical and irrational, or as in the case of the legislator, by design; in other words, either incidental or intentional" (*Elementary Structures of Kinship*, p. 100). This is a travesty of the real views, or "postulates," of the pioneers.

The early scholars who discovered the cross-cousin alliance did not tear it out of its historical context. They provided the starting point for uncovering its origin and meaning. One of Edward B. Tylor's achievements was coining the term "cross-cousin," which signified that it was fundamentally different from our cousin system. He also made a direct connection between cross-cousin marriage and the dual organization, with the formula that a man must not marry a woman of his own *veve* or maternal clan; he must marry a wife on the other side, that is of the woman's *veve* or maternal clan.

Tylor was among the first to correctly point to the fundamental reason for the dual organization and the cross-cousin alliance. He took a different tack from those who believed it was instituted to halt the previous practice of "promiscuity" and "incestuous" marriages between "brothers and sisters." Tylor detected that these institutions were not so much connected with sex or marriage as with the problems created by hostile groups of men incessantly fighting one another.

This is the basis for his oft-quoted statement that for men of that epoch it was a question of "marrying out or being killed out." By itself, however, this does not explain why primitive men fought one another, or the part played by women in creating the alliances that brought these hostile men together in peaceful fraternal relations.

Ignorance of natural death was at the bottom of primitive fighting among men. Just as primitive people did not know how babies were conceived, they did not know the biological facts about death. They believed that anyone who died was "killed" by some enemy. Applied to primitive times, the term "kin" means that anyone who was not of the kin was a non-kin stranger—or enemy. Under the "blood bond" that tied the kinsmen together, the stranger-enemy was the one who brought death to the community and had to be punished.

This led to what is called the "blood revenge" system, which is also referred to as the "blood feud" and "blood debt." In the fights and counterfights which ensued, each side sought to inflict reprisals on the other side to satisfy the "blood debt." As Tylor put it, the avenger of blood was only "doing his part toward saving his people from perishing by deeds of blood" (*Woman's Evolution*, p. 221).

The solution was found by the women of the maternal clans. Through the peace agreements they effected between the groups, the cross-cousin alliances were instituted by which the men of both sides came into a fraternal relationship with each other as male cross-cousins or "brothers-in-law." This in turn permitted peaceful mating relations between the male and female cross-cousins on both sides.

To be sure, since deaths continued, old suspicions and antagonisms were often rekindled and the alliances frequently broke down. This accounts for the hostility that was interwoven with the fraternal relations between the male cross-cousins. It also explains why women at that stage might say "We marry our cross-cousins" or "we marry our enemies." Nevertheless, cross-cousinship grew at the expense of enmity.

As Tylor remarks:

> the intermarrying clans do nevertheless quarrel and fight. Still by binding together a whole community with ties of kinship and affinity, and especially by the peace-making of the women who hold to one clan as sisters and to another as wives, it tends to keep down feuds and to heal them when they arise [ibid., p. 203].

Thus, to see the cross-cousin institution as simply an intermarrying arrangement between men and women is to see only one side of it. On its other and more fundamental side it represents a drastic change in the relationship between men. Through the cross-cousin alliance, non-kin strangers and enemies were converted into new categories of kin and friends—a necessary prelude to matrimonial relations between men and women.

Lévi-Strauss takes up only one side of this cross-cousin institution, the intermarrying arrangement. Upon this one-sided view he constructs the central theme of his book—the eternal manipulation of women by men through "sister exchange." In fact, since the cross-cousin alliance was primarily an exchange of fraternal relations between men, it was far more a system of "brother exchange" than of "sister exchange."

Further testimony of the nature of this "exchange" is that when the first changes of residences took place by the cross-cousins—who formerly lived apart—they were made by the men, not the women. The male cross-cousins left their own maternal clan to become the "visiting husbands" in the maternal clans of their wives. This earliest form of pair-matrimony is called "matrilocal marriage."

These significant facts do not appear in Lévi-Strauss's book. He seems unaware of the magnitude of the blood revenge system, which kept men at odds with one another, and refers to it only in the most superficial sense as a "vendetta" between individual men. According to his male-biased viewpoint, men have always been dominant over women, trafficking in sisters and sisters' daughters. He writes:

> The whole structure of cross-cousin marriage is based on what we might call a fundamental quartet, viz., in the older generation, a brother and a sister, and in the following generation, a son and a daughter, i.e., all told, two men and two women; one man creditor and one man debtor, one woman received and one given [*Elementary Structures of Kinship*, pp. 442-43].

To Lévi-Strauss, the "debt" is not the "blood debt" incurred by men through their blood revenge system. He construes it as a marriage transaction carried out in the modern bookkeeping sense, where debits and credits are incurred in the course of the marketing of products—in this case, women.

Anthropologist Eleanor Leacock, in a probing criticism of the male-centered Lévi-Strauss, shows the absurdity of applying his stereotyped "woman exchange" to matrilineal-matrilocal societies, which are conspicuously egalitarian. As she puts it:

> The terminology of woman exchange distorts the structure of egalitarian societies, where it is a gross contradiction of reality to talk

> of women as in any sense "things" that are exchanged. Women are *exchangers* in such societies, autonomous beings who, in accord with the sexual division of labor, exchange their work and their produce with men and with other women. . . . Lévi-Strauss's formulation mystifies history and the major changes that have taken place in family forms and the relations between women and men ["The Changing Family and Lévi-Strauss, or Whatever Happened to Fathers?" in *Social Research*, summer 1977, pp. 258-59].

But Lévi-Strauss does not recognize the egalitarian structure of primitive society any more than he does its communal economic structure. He becomes exasperated with the early scholars, who, despite the rampant prejudice against women that prevailed in the nineteenth century, did not conceal their discoveries about either the communal economic structure or the esteemed position of women in primitive society. Thus he singles out Frazer, who did make some one-sided and incautious statements about sister exchange, to berate him for sticking to his evolutionary approach. Caustically he writes, "what we regard as the means of escaping from cultural history he attempted to interpret as cultural history" (*Elementary Structures of Kinship*, p. 136).

Indeed, it is only through his own attempted "escape from history"—which no one and nothing can actually escape—that Lévi-Strauss propagates his antiwoman thesis and tries to bury the maternal clan system under a mountain of misinterpreted "patrilineality."

Patrilineal Kinship in the Maternal Clan

Although Lévi-Strauss does not once mention the matriarchy (it is a "dirty word" in most academic circles today), he cannot ignore the matrilineal system of kinship and descent. Indeed, it would be difficult to do so, since this survival of the maternal clan system has persisted in some regions up to our times. But he denies any historical succession from matrikinship to patrikinship.

He not only considers such an evolutionary sequence an "optical illusion" on the part of the early scholars but asserts that "any human group might, in the course of centuries, develop alternately matrilineal or patrilineal characteristics without the most fundamental elements of its structure being profoundly affected by it" (ibid., p. 409).

This is a typical example of Lévi-Strauss's method of mingling half-truths with errors to obfuscate an important issue. It is true that patrilineal kinship did not alter the basic structure of the maternal clan system, but this is so *because it was born in the maternal clan system* and was assimilated into it as a new classification. But it is false to say, as he does, that any human society can flop back and forth haphazardly or at will between kinship systems.

Attempts to conceal the priority of matrilineal kinship go hand in hand with the rejection of the matriarchy. Nevertheless, the record clearly shows that patrikinship (the recognition of a paternal relationship between a woman's child and her husband) came later in history than matrikinship; and the change from one to the other has always been from the matrilineal to the patrilineal, never the other way around (*Woman's Evolution*, pp. 165-69). This historical sequence, in fact, furnishes the starting point for solving certain problems that have persisted with respect to patrilineal kinship and "descent."

Patrikinship evolved out of the cross-cousin mating alliance. As the cross-cousins formed pairing couples, they began to live together under one roof. At first it was the male cross-cousin who took up residence as the "visiting husband" in his wife's locality; this was known as matrilocal marriage. Partly due to his origin as non-kin and partly because he was the stranger in the clan of his wife, he is also referred to as the stranger or foreigner.

In the course of time the wife, too, began to leave her own clan to take up residence in her husband's clan. This is called patrilocal marriage because the wife was residing in her husband's locality. In both instances the change of residence lasted only as long as the marriage, which was usually short-lived, after which the partners returned to their own clans. If the marriage lasted, the pair frequently migrated between the two localities, living part of the time in the wife's locality and part of the time in the husband's locality.

The term "patrilocal marriage," however, led to a profound misunderstanding on the part of many anthropologists that the husband's clan is somehow different from the wife's clan and outside the maternal clan structure. Nothing could be further from the truth. The fact is, whether the wife resides with the husband in his locality or the husband with the wife in her locality, *both are matrilocalities and both belong to the maternal clan structure.*

The same error was made in regard to patrilineal kinship. The mistaken notion grew up among some anthropologists that, just as matrilineal kinship and descent were traced through the mother-line, so patrilineal kinship and "descent" were traced through the father-line. The patrilineal clan seemed somehow different from the matrilineal clan and outside the clan structure. This, too, was a fundamental error.

In reality, there was no such thing as a patrilineal clan existing by itself. *Every clan was structurally a matrilineal clan* belonging to the maternal clan system. It is only in the patrilineal *relationship* between separate maternal clans that we can speak of a "patrilineal clan."

Thus, to the wife whose cross-cousin husband had become recognized as the father of her child, his clan is her "patrilineal clan." But to the husband, his clan is his "matrilineal clan." Similarly in reverse: to the husband's sister, his wife's clan is her (the sister's) "patrilineal clan." But to the sister, her clan is her matrilineal clan, just as it is to her brother. This is the effect of a patrilineal *relationship* superimposed upon an existing matrilineal *structure*.

But there is more to the matter than this. The recognition of the husband as the father of the wife's child *did not include descent through the father-line*. One of the gravest errors made in anthropology has been to suppose that wherever patrilineal kinship was found, it was accompanied by patrilineal descent. This is not the case. The original form of male descent, as a line of descent subsidiary to the matrilineal line, was *male descent through the mother's brothers*. This continued on even after patrilineal kinship was recognized.

This thesis of original male descent through the mothers' brothers represents a breakthrough in solving the most perplexing problems of patrilineal kinship, and it was set forth for the first time in *Woman's Evolution*. Among other things, it explains the continuing preeminent position of the mother's brother over the mother's husband even after patrilineal kinship was recognized. And it also discloses the source of the antagonisms and conflicts between the brothers and husbands.

These conflicts developed because only the matrilineal kin were the "blood kin," tied together by the "blood bond." The patrilineal kin were excluded. This made it extremely difficult to change the line of descent from the mother's brother-line to the husband's line, since the "blood bond" was invoked whenever deaths or injuries occurred to the matrilineal kin. Neither the cross-cousin alliance nor the recognition of patrilineal kinship could prevent these conflicts between the matrilineal kinsmen and the non-kin strangers and enemies—the men who had to pay their "blood debt."

These conflicts between brothers and husbands invaded the little pairing family, which became essentially a "divided family." The same men who were in conflict with each other through the "blood feud" were also performing double fatherhood functions for the children of the family; on the one side the mother's brother as the former "father" of his sister's children, on the other side the mother's husband as the father of his wife's children. This created intolerable contradictions and tensions within the family and in society at large.

Thus it became imperative to get rid of the one-sided matrilineal kinship system with its "blood" obligations, to eliminate the mother's brother, and to change the line of descent to the father-line to conform with the recognition of patrilineal kinship. The "divided family" became unified once it was transformed into a one-father, or patriarchal family. How this was accomplished, first through the bloody travail of sacrifice and then through the rise of private property, is spelled out in *Woman's Evolution*. The same process that eliminated the mothers' brothers and gave birth to the patriarchal family also brought down the whole maternal clan system, replacing it with a totally new social system—patriarchal class society.

The early scholars, grappling with these extremely difficult problems of a changing kinship system, made considerable headway, despite their errors. But with the turn away from their evolutionary method, these errors were carried forward and deepened by the antihistorical anthropologists. They juxtaposed the patrilineal clan against the matrilineal as a totally distinct clan. Then they put all the emphasis on the side of the patrilineal clan, to the extent that it soon became indistinguishable from a large father-family. By and by, the maternal clan was shoved out of the picture altogether, and the father-family, with father-descent, took its place. The patrilineal clan, patrilineal kinship and descent, patrilocal marriage, and patrilocal residence became all-pervasive, with only some vague references to an aberrant and unimportant custom called "matrilineal kinship and descent."

Lévi-Strauss's "elementary structures of kinship" is one of the most bizarre compilations of these errors. Having rejected the archaic period of history along with the maternal clan system, he does not understand that the large clan categories of kinship relationships cannot be crammed into the little circle of father-family relatives. He flounders around trying to explain puzzling phenomena that do not fit into his thesis of the eternal father-family.

For example, he takes an early formula, that "a man marries his mother's brother's daughter" (which was merely a clumsy way of expressing cross-cousin marriage), and blows it up into an unrecognizable edifice of "marriage with the father's sister's daughter" and "patrilateral cousin marriage," etc. He has gone so far in these respects that he has embarrassed even some of his colleagues. Cross-cousin marriage belongs to the maternal clan system and existed before the father and father's sister made their appearance in history.

Again, on the question of patrilocal marriage, Lévi-Strauss is bewildered by the fact that

> matrilineal descent is accompanied by patrilocal residence. . . . The husband is a stranger, "a man from outside," sometimes an enemy, and yet the woman goes away to live with him in his village to bear children who will never be his. The conjugal family is broken and rebroken incessantly. How can the mind conceive of such a situation? How can it have been devised and established? [*Elementary Structures of Kinship,* p. 116.]

His answer is characteristic of his male-biased viewpoint. He writes,

> It cannot be understood without being seen as the result of the permanent conflict between the group giving the woman and the group acquiring her. Each group gains the victory in turn, or according to whether matrilineal descent or patrilineal descent is practised. The woman is never anything more than the symbol of her lineage. Matrilineal descent is the authority of the woman's father or brother extended to the brother-in-law's village [ibid., p. 116].

Apart from his other distortions, Lévi-Strauss refuses to examine the real causes of the conflicts between groups of men, i.e., the demands created by the "blood bond" and the "blood debt." He prefers to look in the direction of women and make them, as objects or property, the cause of the conflicts. He resorts to the same false assertion propagated by others that women have always been under the dominion of men; if not their fathers, then their brothers. By foreclosing history, the mothers' brothers—whose authority was second to that of the mothers and sisters—are also smuggled into this eternal male domination over women.

Despite Lévi-Strauss's antipathy to the historical approach, he prefers not to proclaim it openly but rather to gloss over it. For example, he presents his three "elementary" structures of marriage in the following order:

Bilateral marriage
Patrilateral marriage
Matrilateral marriage

He expresses concern that this arrangement might be taken for a historical sequence and asks rhetorically: "Does this logical priority correspond to an historical privilege?" Disavowing this, he contends that he is confining himself only to a "structural analysis." Dodging the issue of historicity, he remarks, "It is for the cultural historian to inquire into this" (ibid., p. 465).

Such a "cultural historian" would have to acknowledge that a historical sequence is involved. The progression goes from matrilocal to patrilocal marriage, both within the framework of the maternal clan system. And, no matter what Lévi-Strauss means by his hybrid term "bilateral marriage," the next stage in the evolution of marriage came after the social overturn that brought the father-family into existence, and along with it the patriarchal system of marriage.

Lévi-Strauss's exposition on the distinctions between family and clan kinship is no more coherent than his picture of marriage forms. He writes:

> For a long time sociologists have considered that there is a difference in nature between the family, as found in modern society, and the kinship groups of primitive societies, viz., clans, phratries and moieties. The family recognizes descent through both the maternal and paternal lines, while the clan or moiety reckons kinship in one line only, either the father's or the mother's. Descent is then said to be patrilineal or matrilineal. . . . A system with matrilineal descent does not recognize any social kinship link between a child and its father, and in his wife's clan, to which his children belong, the father is himself a "visitor," a "man from outside," or a "stranger." The reverse situation prevails in a system with patrilineal descent [ibid., p. 103].

Here again truth is mingled with error. Lévi-Strauss makes no distinction between kinship and descent, using the terms loosely and interchangeably. He presents two separate "unilineal" lines of descent, one through the father-line and the other through the mother-line. Then he states, speaking of matrilineal descent, where no kinship link is recognized between a child and its father, that "the reverse situation prevails in a system with patrilineal descent."

It is certainly true that, before patrilineal kinship made its appearance in history, no paternal link was recognized between father and child—much less a descent line. But in what society, however primitive, has the kinship and descent line between mother and child not been recognized? Lévi-Strauss gives no illustration. Instead he shifts to a criticism of "certain writers" (unnamed) who have accepted these "unnatural customs" as the "true picture," and then he proceeds to correct the picture.

This comes down to defending the proposition that patrilineal kinship and descent have always been recognized, even though this contradicts his own previous statement. He writes, "Today we know that societies as matrilineal as the Hopi take account of the father and his line, and that this is so in most cases" (ibid., p. 104). By inference, then, the family and bilineal kinship have always existed.

If this is so, what becomes of the two separate unilineal lines of kinship and descent that he has set forth, the one through the father-line and the other through the mother-line? Weaving our way through some convoluted discourse, we learn the following:

> There are certainly no watertight bulkheads between unilineal, bilineal and undifferentiated descent. Any system has the coefficient of diffuse nondifferentiation resulting from the universality of the conjugal family. To a certain extent, moreover, a unilineal system always recognizes the existence of the other line. Conversely, it is rare to come across an example of strictly undifferentiated descent. Our society, which has gone far in this direction (inheritance comes equally from the father and the mother, and social status is received, and prestige derived, from both lines, etc.), keeps a patrilineal inflection in the mode of transmitting the family name. Be that as it may, the importance of undifferentiated systems for anthropological theory is today unquestionable [ibid., p. 106].

Through a sleight of hand, Lévi-Strauss's two separate lines of unilineal descent are now fused and smuggled into the bilineal family system of kinship and descent. This is accomplished by means of a hitherto unknown anthropological category—

"undifferentiated" descent. He doesn't tell us what this term signifies. But elsewhere he writes that

> it is necessary to determine a mode of descent, which can be undifferentiated, unilateral (here, either patrilineal or matrilineal), or bilateral (i.e., in which each individual will be given both a patrilineal reference and a matrilineal reference) [ibid., p. 323].

The question arises: how does his new term, "undifferentiated," differ from the old term of "bilateral" descent, if both are reckoned through the father and mother? He doesn't say until eighteen years later, in his preface to the second edition. Complaining about the criticisms directed against him on this and other obfuscations, even though he consulted more than seven thousand books and articles, he writes:

> I might have indicated the importance of a general study of so-called "bilateral" or "undifferentiated" systems of descent, even though I did not undertake such a study. These systems are far more numerous than was believed when I wrote my book, although, by a natural reaction, there has been perhaps too great a haste to include, in these new types, systems which are now being seen, more and more, as possibly reducible to unilateral forms [ibid., p. xxviii].

This garbled apology still does not tell us what, if any, difference there is between the "two" systems of "undifferentiated" and "bilateral" descent. But it does reveal something about the shifty and often unintelligible methods of analysis which permeate his book.

Lévi-Strauss's objective is to bury the maternal clan kinship structure inside the father-family structure in order to perpetuate the myth that patriarchal class society, the father-family, and male supremacy have always existed—and along with these the exchange and manipulation of women by men. To buttress this thesis, he brings forth his "incest" fantasy.

Lévi-Strauss's "Incest" Fantasy

Lévi-Strauss subscribes to the doctrine that because men wield social and political power today, they represent the superior sex which has always dominated women. He writes that

> It is because political authority, or simply social authority, always belongs to men, and because this masculine priority appears

> constant, that it adapts itself to a bilineal or matrilineal form of descent in most primitive societies, or imposes its model on all aspects of social life, as is the case in more developed groups [ibid., p. 117].

Actually, since the women of the maternal clan period were the prime producers of the necessities of life, they held the highest authority as the social and cultural leaders, a fact that is exhaustively documented in *Woman's Evolution.* Lévi-Strauss, who doesn't deal with the economics of primitive society, tries to prove eternal male supremacy through his thesis on "sister exchange."

Concealing the fact that in cross-cousin matrimony there is just as much "brother exchange" as "sister exchange," he writes, "the basic fact [is] that it is men who exchange women, and not vice versa." And he adds:

> The total relationship of exchange which constitutes marriage is not established between a man and a woman, where each owes and receives something, but between two groups of men, and the woman figures only as one of the objects in the exchange, not as one of the partners between whom the exchange takes place [ibid., p. 115].

Here again, what is true is intermingled with what is false. It is true that only men were "partners" in the exchanges that were made between them. That is because they, not the women, had to solve the problem of their hostility toward one another. This was solved by making them "partners" in the exchange relations.

From the female side, far from being "objects" exchanged by men, it was the women who made the matrimonial alliances between the groups just as they made the peace agreements between the men of the groups. And, as the anthropological record shows, it was the woman, not the man, who "made the proposal of marriage," and they made it of their own free choice (*Woman's Evolution*, p. 310).

How, then, does Lévi-Strauss arrive at his completely wrong stance?

His theory on woman-exchange stems from the primitive custom called "gift-giving" or "gift-exchange." He borrows from Marcel Mauss's famous essay, *The Gift*, published first in French in 1925 and in English in 1967. But, here again, Lévi-Strauss takes a one-sided view of the institution, misinterpreting his source material to arrive at the conclusion congenial to him, that women have always been exchangeable objects in the hands of men. Let us examine his distortions.

First, the gift-exchange system, or "reciprocity" as it is also called, was exclusively a feature of the archaic maternal clan period. It was characteristic of that collectivist, egalitarian society with a low level of productivity. There was no element of purchase or sale, or even of barter, in the system of gift-reciprocity; such economic practices had not yet come into existence. Gift-giving therefore can be called a system of "interchange" to distinguish it clearly from its opposite, the modern generalized system of commodity exchange, where everything necessary for life is bought and sold.

Second, the primitive interchange system grew up out of the need to solve the problems of men. That is, to overcome the fear and hostility that separated kinsmen from non-kin strangers and enemies and bring them together as allies and friends. Thus, while modern commodity exchange and commercialism, based on private ownership and fostering individualism, alienates men from one another, the primitive gift-interchange system brought hostile or potentially hostile men together as new kinds of kin and cross-cousins. (See *Woman's Evolution*, pp. 211-71.)

Large gift-giving festivals were regularly held, and everything necessary for peaceful social life was interchanged at these festivals—from food and sex, to crafts, courtesies, dances, and rituals. Similar festivals were held on occasions of "life crises" such as births, deaths, matrimonies, and peace pacts. They were called *potlatches* in some regions and *corrobborees* in others. Whole clans, phratries, and even tribes participated in these get-togethers to establish or confirm fraternal alliances.

Lévi-Strauss reduces this massive tribal interchange system down to individual family men trafficking in women in their marriage exchange market. He writes that "two types of men always have virtual authority over the same woman, viz., her father and her brother," and adds:

> Therefore we immediately have four possible schemes of reciprocity: the exchange of sisters by or for the brothers, either in the older generation or in the younger generation; the exchange of daughters by fathers; and finally, the exchange of a sister against a daughter, or of a daughter against a sister. Furthermore, the four schemes can coexist, a man not being restricted to only one daughter or to only one sister, but being able to have several of them with nothing to prevent them from being exchanged according to different modalities [*Elementary Structures of Kinship*, p. 433].

Apart from the fact that this patriarchal harem-type situation

A festival of the Virginia Algonquin tribes before the colonists settled in America. Women and men dance in a circle enclosed by carved posts depicting women's heads. As dancers tire they drop out and others take their places. In the center of the ring, three women—their arms around one another in a manner resembling the Three Graces—turn round and round as the dance proceeds. In Indian life the Three Sisters of vegetation were known as Bean, Corn, and Squash. Women played an important part in Indian festivals and corroborees, which established and reaffirmed peace and fraternity among the men of different groups and tribes.

Theodore De Bry, 1590

did not exist in the period of the maternal clan, the question arises: how and when did men get the power to traffic in women? Historical materialists have explained that such male supremacy began only a few thousand years ago—with the overturn of the maternal clan system by patriarchal class society. But Lévi-Strauss has a different outlook. He believes that the exchange of women by men goes all the way back to the beginning of human life (about a million years ago), and it came about through the operation of what is to him the most fundamental law in all history—the incest taboo!

The "incest theory" to explain the primitive taboo is an old theory. Incest is defined as sexual relations between the closest genetic relatives. This incest theory was originally based upon the mistaken notion that clan brothers and sisters were such close genetic relatives that inbreeding among them would be harmful to the species. Subsequently, with the advances made in biological knowledge about the effects of inbreeding, this argument was undermined. In addition, it was found that classificatory kin were in the overwhelming majority not genetically related at all. This further undermined the biological basis for the incest theory.

It should be apparent that to undermine or discard the biological basis for the taboo is to remove the "incest" from the prohibition. Whatever the taboo was directed against, it could not, then, be characterized as an "incest taboo." This problem created the need for some new theory to take the place of the incest theory. Unfortunately, the "new" theories were by and large attempts to make the untenable incest theory more tenable. Psychological reasons were advanced to take the place of the discredited biological reason. But these could not be sustained under rigorous scrutiny either.

The most implausible aspect of the incest theory was the premise that primitive people, who did not even know the biological facts about birth and death, could have understood anything about incest. Such knowledge requires a very high scientific level of understanding and was acquired only in the most recent period of social development.

The "incest taboo" thus became the Number One Baffler in anthropology. Three options remained open after repeated setbacks in trying to fathom its meaning: to continue to believe in a biological basis for the prohibition; to stretch the definition of "incest" to include categories of "fictional" relatives who are not genetic relatives; or, barring these two devices, to abandon the search altogether as an impenetrable mystery.

Lévi-Strauss holds an ambiguous and contradictory position on the subject. He dismisses the idea that primitive people could have known anything about incest since, as he writes, such knowledge dates only from about the sixteenth century (p. 13). And he is equally skeptical about the psychological reasons advanced in place of the biological. What Lévi-Strauss seeks is a sound *sociological* explanation for so all-pervasive a phenomenon as the taboo.

Unable to produce such an explanation, however, Lévi-Strauss winds up with only a new variation of the incest theory. He then bedecks this old theory in new raiment. Despite all his uncertainties and ruminations on the subject, Lèvi-Strauss is certain of one thing: the incest taboo is the oldest prohibition in history and represents the dividing line between nature and culture. He writes:

> Here therefore is a phenomenon which has the distinctive characteristics both of nature and of its theoretical contradiction, culture. The prohibition of incest has the universality of bent and instinct, and the coercive character of law and institution. Where then does it come from, and what is its place and significance? Inevitably extending beyond the historical and geographical limits of culture, and co-extensive with the biological species, the prohibition of incest. . . . presents a formidable mystery to sociological thought [ibid., p. 10].

It is somewhat illogical for Lévi-Strauss, who rejects the archaic period as an illusion, to assert that there is one law that has prevailed for all time—and that is the incest prohibition. Under the circumstances, how does he know this? Nevertheless he writes, "The incest prohibition is at once on the threshold of culture, in culture, and in one sense, as we shall try to show, culture itself" (ibid., p. 12).

Lévi-Strauss's placement of the incest prohibition at the very center of human society and culture flows from his thesis that the father-family is eternal. In his view the incest prohibition was designed to prevent consanguine marriages within the family. These consanguine marriages, even if not biologically harmful (he is not quite sure that they were not), were socially harmful in keeping the tiny family unit separated from other little family units and in conflict with one another. He writes:

> If these consanguine marriages were resorted to persistently, or even overfrequently, they would not take long to "fragment" the social group into a multitude of families, forming so many closed systems or sealed monads which no pre-established harmony could prevent from proliferating or from coming into conflict [ibid., p. 479].

Therefore, through sister exchange the little biological unit of father-mother-child was saved from further fragmentation through consanguine marriages.

This circulation of women among men represents the "positive" feature of the incest prohibition, according to Lévi-Strauss. Summing up, he says, "Considered from the most general viewpoint, the incest prohibition expresses the transition from the natural fact of consanguinity to the cultural fact of alliance" (ibid., p. 30). At the end of his book he emphasizes that it is not a prohibition like any others, "It is *the* prohibition" (ibid., p. 493).

Casting about for some authority to buttress his incest theory, Lévi-Strauss suggests that it was also Tylor's view. He writes that some anthropologists, "following Tylor, found the origin of the incest prohibition in its positive implications." To this he adds, "As one observer rightly puts it, 'An incestuous couple as well as a stingy family automatically detaches itself from the give-and-take pattern of tribal existence; it is a foreign body—or at least an inactive one—in the body social'" (ibid., p. 488).

It is Lévi-Strauss, not Tylor, who reduces the tribal interchange system, designed to bring about fraternal relations between hostile groups of men, to a stingy little family circle. Tylor's writings on the alliances made in the primitive period of the maternal clan were not hinged on any "positive implications" of the incest prohibition. In fact, he was one of the few who avoided falling into the trap of accepting the incest theory.

What Tylor saw were the dangers involved in fragmenting the network of clans, phratries, and tribes into warring groups of men—not the danger of incest in a little family circle as Lévi-Strauss implies. Thus what Tylor emphasized was the peacemaking of the women, which wove together this fraternal network through kinship affiliations and alliances—and held it together until society could progress to a higher stage.

In other words, men did not become cultured human beings, demarcated from the primates, through any "positive implications" of a nonexistent "incest prohibition." Our species was acculturated through the work and wisdom of the women.

Lévi-Strauss's incest theory is no more tenable than the others he rejected. Yet he is scornful of those who gave up the struggle to find a rational explanation for the "incest" taboo. He castigates "contemporary sociology" because it "has often preferred to confess itself powerless than to persist in what, because of so many failures, seems to be a closed issue. . . . It declares that the prohibition of incest is outside its field" (ibid., p. 23). But these sociologists are at least more cautious than Lévi-Strauss, who insists on elevating a nonexistent incest taboo as the keystone of human culture.

What was the origin and aim of the taboo? The first new theory that replaces the erroneous incest theory can be found in *Woman's Evolution.* It presents a logical and documented theory that the primitive taboo was a double taboo directed against the menace of cannibalism and against any return to animal sexual violence. Neither of these taboo clauses have anything whatever to do with incest. This taboo riddle, however, could only be solved through a scrupulous adherence to the historical method, an unbiased view of women, and a recognition of the priority of the maternal clan system.

Lévi-Strauss wants us to believe that humanity began not with the maternal clan but with the patriarchal family and male domination over women. The result is a grotesque depiction of the "incest prohibition" as the cornerstone of society and culture. He also wants us to believe that he has no prejudice against women. For example, he refers to them as the "most precious" of "gifts." A closer look, however, shows that he does not at all understand the primitive gift-giving system. And he regards women not so much as gifts but as "goods" to be merchandized like any other products.

Lévi-Strauss's Marriage Market

Like everything else in primitive life, the gift-giving, or interchange, system evolved and changed as it passed from a lower to a higher stage of development. In the earlier form of cross-cousin exchanges, two moieties or phratries of one tribe exchanged fraternal relations between the men and matrimonial relations between the men and women of both sides.

In the course of time, with the advent of intertribal exchange relations, the simple moiety system expanded outward in a more complex and "circular" system of interchanges. One of the best illustrations can be found in Malinowski's description of the *kula* in the Trobriand Islands, where whole tribes participated in a massive interchange circuit. This represents a survival of the last stage of the primitive interchange system before tribal society and its gift-giving were overturned and replaced by patriarchal class society and its system of commodity exchange.

Lévi-Strauss does not deal with these evolutionary sequences of the gift-giving system or with the drastic change that occurred when it was replaced by commodity exchange. Instead he conceals this evolution and revolution by introducing two terms: "restricted exchange" and "generalized exchange." Wading through his complicated discourses, it seems that the restricted is the more archaic form, the generalized the later form. But, with his antipathy to history, Lévi-Strauss takes refuge behind the "structuralism" to which he is limiting himself. He writes:

> What is the connexion between restricted and generalized exchange? Are they to be seen as two independent forms, yet capable of interacting one upon the other when the chances of culture contact bring them together, or do they represent two related stages in one process of evolution? In so far as the concern is with the solution of regional problems, this is a problem for the ethnographer and cultural historian. Our own intention is limited to making a structural study of the two types and their interrelations. . . . [Ibid., p. 220.]

His structural study comes down to an unchanging commodity-exchange system as we know it today, with a little gift-giving occasionally thrown in (similar to the way we exchange gifts at Christmas time). He cannot conceive of a period of time when there was no other kind of exchange relationship than gift interchange, just as he cannot conceive of a period of time when social and sexual equality prevailed. He states, parroting the current line, that sometimes in the primitive system the "profit" from gift-exchange may be delayed or there may be an indirect kind of remuneration, such as gaining prestige, status, or future favors.

He gives as an example the *potlatch* of the Northwest Coast Indians where certain intangibles such as prestige and authority are acquired along with "an appropriate amount of interest, sometimes as much as 100 per cent." In other instances "the profit is neither direct nor inherent in the things exchanged," because there is "clearly something else in what we call a 'commodity' that makes it profitable to its owner or trader. Goods are not only economic commodities, but vehicles and instruments for realities of another order, such as power, influence, sympathy, status and emotion. . . ." Elsewhere, "A gift is at most a venture, a hopeful speculation" (ibid., pp. 52-55).

Thus Lévi-Strauss equates the primitive gift-giving system, designed to maintain peace and fraternity among clans and tribes, to modern stores in Latin countries called "casas de regalias" and Anglo-Saxon "gift shops," where free food and drink are served to the customers buying the gifts.

Anticipating the criticism that this travesty of the primitive gift-giving system will arouse, Lévi-Strauss writes:

> Perhaps we shall be criticized on the ground of having brought together two dissimilar phenomena, and we will answer this criticism before proceeding. Admittedly, the gift is a primitive form of exchange, but it has in fact disappeared in favour of exchange for profit, except for a few survivals such as invitations, celebrations and gifts which are given an exaggerated importance. In our society, the proportion of goods transferred according to these archaic modalities is very small in comparison with those involved in commerce and merchandising. Reciprocal gifts are diverting survivals which engage the curiosity of the antiquarian; but it is not possible to derive an institution such as the prohibition of incest, which is as general and important in our society as in any other, from a type of phenomenon which today is abnormal and exceptional and of pure anecdotal interest. [Ibid., p. 61.]

It is certainly true that Lévi-Strauss's fantasy about the incest prohibition cannot be derived from the primitive form of exchange. But it is not true that survivals of the gift-giving system are merely diversions for antiquarians, or that they are "abnormal" or only of "anecdotal interest."

The gift-interchange institution is part and parcel of the maternal clan system that had endured for the greater portion of human life on earth, compared to the few thousand years of class society and its system of commodity exchange. Survivals of the archaic gift-exchange system point the way back into prehistory, when society was egalitarian and when women occupied the leading place in social and cultural life.

The fact that Lévi-Strauss ignores the egalitarian character of primitive society goes hand in hand with his obliteration of the matriarchy. His views run counter to the findings of the pioneer anthropologists that private property and class divisions did not exist in tribal society. By the same token, there were no purchases or sales of commodities in the gift-interchange system. While Lévi-Strauss has almost nothing to say on the socioeconomic structure of primitive society, he does take a slap at Frazer because Frazer "pictures an abstract individual with an economic awareness, and he then takes him back through the ages to a distant time when there were neither riches nor means of payments" (ibid., p. 139).

To Lévi-Strauss, men have always paid for their property, if not with money, then by barter—women being among the "valuable" objects bartered. To back this up he cites R. Firth, who wrote that, along with ordinary exchanges, "'goods of unique quality'" are also handed over. "'Such for instance was the transfer of women by the man who could not otherwise pay for his canoe. Transfers of land might be put into the same category. Women and land are given in satisfaction of unique obligation'" (ibid., p. 61).

Apart from the dubious assertion that a primitive man traded some women for a canoe, it is deceitful to make an amalgam between this and the land transfers, which were made in a much later period. Those transfers accompanied a marriage-merger between aristocratic land-owners; the lady who married the lord went along with the land. Such transactions belong to patriarchal class society—not to the period of the maternal clan system.

But to Lévi-Strauss, exchange is exchange, and it is one and the same kind of exchange throughout all history. As he puts it, marriage "is always a system of exchange." He gives a large number of variations of this marriage-exchange: it may be a cash or short-term transaction, or a long-term transaction; it may be direct or indirect, implicit or explicit, opened or closed; "secured by a sort of mortgage on reserved categories," or not so secured, etc. Then he winds up:

> But no matter what form it takes, whether direct or indirect, general or special, immediate or deferred, explicit or implicit, closed or open, concrete or symbolic, it is exchange, always exchange, that emerges as the fundamental and common basis of all modalities of the institution of marriage [ibid., p. 478-79].

This is Lévi-Strauss's fanciful marriage market through which men have exchanged women from time immemorial. Trying to disguise his low opinion of women, he sometimes refers to them as the "gift" or the "supreme gift." But fundamentally their value lies in their marketable services and usefulness to men.

Women are "valuables *par excellence* from both the biological and the social points of view," he writes. And, "The prohibition of incest is less a rule prohibiting marriage with the mother, sister or daughter, than a rule obliging the mother, sister or daughter to be given to others. It is the supreme rule of the gift. . . ." (Ibid., p. 481.) The "gift" is not a woman giving herself to a man of her own free will but a man transacting with another man the exchange of the woman. To Lévi-Strauss, woman-exchange in the marriage market "tends to ensure the total and continuous circulation of the group's most important assets, its wives and daughters" (ibid., p. 479).

To establish the marketability of these "assets" Lévi-Strauss sometimes resorts to innuendo. He writes, "There is no need to call upon the matrimonial vocabulary of Great Russia, where the groom was called the 'merchant' and the bride, the 'merchandise' for the likening of women to commodities, not only scarce but essential to the life of the group, to be acknowledged." And in a footnote, citing Kowalewsky, he adds, "The same symbolism is to be found among the Christians of Mosul where the marriage proposal is stylized, 'Have you any merchandise to sell us?'" (Ibid., p. 36.)

It is true that in patriarchal society, including Greece and Rome of the "civilized" era, men bought their wives from other men and sold their daughters as wives for other men. But it is wrong to say that such transactions took place in primitive society.

Despite his attempts to appear benevolently inclined toward women, Lévi-Strauss does not conceal his disdain for them. Lest women get an inflated sense of their worth because they are "valuable assets," Lévi-Strauss sets them straight. In a discourse on women as a "scarce commodity," he attributes their scarcity to the "deep polygamous tendency, which exists among all men"—and may even stem from the polygamous apes. In any case, this "always makes the number of available women seem insufficient."

He adds, "even if there were as many women as men, these women would not all be equally desirable." In parentheses he cites Hume: "(as Hume has judiciously remarked in a celebrated essay), the most desirable women must form a minority" (ibid., p. 36). Valuable as women may be in the marriage market, to the misogynists only a few are really desirable as women.

The trend of contemporary science, from astrophysics to

zoology, is to trace the development of all things from their origins, through their successive forms, to what exists today. By setting aside historical considerations in presenting his "elementary structures of kinship," Lévi-Strauss is as scientifically backward as a biologist who ignores evolution after Darwin. In this he is not alone.

Lévi-Strauss, the "structuralist," shares the same antihistorical outlook as Franz Boas, R. H. Lowie, A. R. Radcliffe-Brown, and others of his "functionalist" predecessors. They are all united on one central issue: concealing the matriarchal epoch, which discloses the preeminence of women in that first long period of history. However, Lévi-Strauss has gone further than the others. He obliterates the "elementary structures of kinship" which belong to that period—while presenting his treatise under that very title.

Evolutionism and Antievolutionism (1957)

What is the state of anthropology and the main direction of its development in the English-speaking world? How and why have the predominant contemporary schools diverged from the methods used by such pioneers as Lewis Morgan in the United States and Edward B. Tylor in England? These two men were instrumental in establishing the science of anthropology in the second half of the nineteenth century and inspired its first far-reaching achievements. Have the modern academic anthropologists advanced beyond the Morgan-Tylor school, as they claim, rendering the earlier procedures and findings obsolete? Have the Marxist analyses and conclusions regarding ancient society, which relied upon materials provided by these nineteenth-century originators of scientific anthropology, become invalidated?

These questions were posed with special force in a volume of about one thousand pages called *Anthropology Today*. This "encyclopedic inventory," published in 1953 by the University of Chicago Press, resulted from a conference sponsored by the Wenner-Gren Foundation of Anthropological Research. Prepared under the supervision of A. L. Kroeber, dean of the modern American school, it contains fifty inventory papers by "eminent scholars from every continent in the world" and represents "the first great stocktaking of the whole of our knowledge of man as it is embodied in the work of modern anthropologists." It has been supplemented by a second volume, *An Appraisal of Anthropology Today*, which contains critical comments by eighty scholars on the problems posed in these papers and by the state of their science.

The compilation surveys and summarizes such diverse yet related branches of social science as biology, archaeology, anthropology, genetics, linguistics, art, folklore, psychology, and includes techniques of field study and applied anthropology in medicine, government, etc. It is undeniably a useful source and reference book. But it is most instructive and important as a guide to the current methods used by professional anthropologists, disclosing in detail how these scholars systematically approach the basic problems in the study of ancient society and primitive life.

The contributors display a wide variety of nuances in their specific procedures and have many unresolved differences among themselves on this or that aspect of their specialities. This is normal and fruitful. But, with rare exception, they resist any consistently evolutionary method of thought or materialist interpretation of history. This throws them into opposition not only to Marxist historical materialism but to the founders of their own science, the classical school of the nineteenth century.

This represents a profound theoretical reversal in the historical development of anthropology and therefore merits serious examination. One virtue of the Wenner-Gren compilation is that it provides in a single volume abundant materials for such a study. It makes clear how sharp the break is between the nineteenth-century and twentieth-century schools of anthropology, in that the second stage stands in avowed opposition to the premises of the first. It further illustrates the specific nature of the differences separating them.

Since we are dealing with the history of this branch of science, it is necessary to go back to its beginnings to get at the roots and reasons for this division and reversal.

Birth Pangs

Anthropology, like everything else in this world, was born in and through struggle. It emerged as a branch of science about a hundred years ago through a series of colossal battles against religious dogmas and petrified ideas.

The first major dispute centered around the antiquity of humankind. Theologians had established the duration of humanity in accord with the Bible at some six thousand years. Even the great French biologist Cuvier adhered to this orthodox view and argued that fossilized bones of men antedating this time did not exist. However, another Frenchman, Boucher de Perthes, exploded this prejudice by his discoveries of ancient stone axes in French deposits which paleontological tests proved were much older. His book published in 1846, demonstrating that fossil men and their tools dated back tens of thousands of years, was greeted with skepticism and scorn.

Continued discoveries of ancient human fossils and tools soon settled this question beyond dispute. Today, through the findings of paleontology and archaeology, such relics of ancient humanity have been chronologically arranged in time sequences which thrust back the age of humankind to a million years or more. Mysticism in this field was figuratively crushed by the material weight of the bones and stones of ancient humanity.

The second great battle was waged around the animal origin of humankind. It began with the publication in 1859 of Darwin's *Origin of Species*, followed in 1871 by his *Descent of Man*. Darwin's proof that humanity arose out of the animal world, more specifically out of the anthropoid species, was a direct blow to the Adam-and-Eve myth. This was a more serious challenge to the divine origin of humanity than simply pushing the birth of mankind farther back in time. Yet, despite the hostility it encountered, Darwin's view became the point of departure for the first scientific study of the formation of humanity. A biologist applying materialist methods had cleared the road for linking anthropology to natural science.

Darwin confined his studies primarily to the biological preconditions for the emergence of humankind. The study of humankind, however, is predominantly a social study. The science of anthropology therefore began at a much higher rung in the ladder of evolution, with the investigation of primitive peoples in areas remote from civilized centers. By examining these living survivals of primitive society, early anthropologists sought to single out the distinctive features which marked off ancient society from our own; they came up with some very surprising conclusions.

The third major struggle unfolded over two interrelated basic distinctions between the institutions of modern and primitive society: the question of the matriarchy versus the patriarchy, and the question of the clan versus the family. In his book *Das Mutterrecht*, published in 1861, Bachofen, using literary sources as evidence, set forth the proposition that an epoch of matriarchy had preceded the patriarchal form with which we are so familiar. Bachofen noted that one of the most striking features of primitive life was the high social status and exceptional authority enjoyed by primitive women in contrast to their inferior status in the subsequent patriarchal epoch. He believed that this epoch of "mother right" which preceded "father right" resulted from the fact that fathers were unknown and the primitive group identified themselves exclusively through the maternal line.

The question of matriarchy was inextricably linked with the

clan group of primitive times as contrasted with the individual family of modern times. Lewis Morgan, in his book *Ancient Society*, published in 1877, disclosed that the unit of primitive society was not the individual family but the *gens*, or clan.

Engels believed that Morgan's discovery was as important to the study of the primitive social structure as the discovery of the cell was to biology, or as Marx's concept of surplus value was to economics. Given the unit of the gens, or clan, the road was opened for anthropologists to investigate and reconstruct the formation and organization of tribal life. As a result of his pioneer work, Morgan is hailed as the founder of American anthropology.

Morgan believed that the family, as it is constituted today, did not exist in ancient society and is essentially a product of civilized conditions. Before the family came the clan, which was composed not of fathers and mothers but of kinsmen and kinswomen, or clan "brothers and sisters." Morgan also indicated that the clan structure was matriarchal. Thus the dispute around the historical priority of the matriarchy over the patriarchy became inseparable from the controversy around the historical priority of the clan over the individual family.

The fourth and most persistent struggle unfolded around the sharp contrast between the basic economic and social relations of primitive and civilized society. Morgan demonstrated that modern society, founded upon the private ownership of the means of production and divided by class antagonisms between the propertied and nonpropertied, is the opposite of the way primitive society was organized. In the primitive community, the means of production were communally owned and the fruits of their labor equally shared. The clan was a genuine collective in which every individual was provided for and protected by the entire community, from the cradle to the grave.

This most basic feature of primitive life was characterized by Morgan and Engels as "primitive communism." But this collectivist social system as well as its matriarchal aspects were frowned upon and discounted by those who wished to perpetuate the dogma that the modern system of private property and class distinctions have persisted without essential change throughout the whole history of humankind.

The struggles around these four major issues, which arose through the researches of the nineteenth-century pioneer thinkers, gave birth to the science of anthropology. Although many questions remained unanswered, the classical school of anthropologists provided the keys for opening a series of hitherto closed doors into the recesses of ancient society. They were founders of the scientific investigation into prehistory.

The Classical School

The twin stars of anthropology in the English-speaking world in the latter part of the nineteenth century were Morgan in the United States and Tylor in England. Around them were a galaxy of brilliant scholars and field workers who made noteworthy contributions to various aspects of this science. Their work was, of course, supplemented by equally able workers in Europe and in other countries.

The work of this pioneer school was marked by the following traits: It was, first of all, evolutionary in its approach to the problems of precivilized humanity. These anthropologists extended Darwinism into the social world. They proceeded on the premise that in its march from animality to civilization, humankind had passed through a sequence of distinct, materially conditioned stages. They believed that it was both possible and necessary to distinguish the lower stages from the higher ones that grew out of them and to trace the interconnections between them.

Secondly, this school was substantially materialist. Its members laid great stress upon the activities of human beings in procuring the necessities of life as the foundation for explaining all other social phenomena, institutions, and culture. They sought to correlate natural conditions, technology, and economics with the beliefs, practices, ideas, and institutions of primitive peoples. They probed for the material factors at work within society to explain the succession and connection of different levels of social organization. The most successful exponent of this evolutionary and materialist method was Morgan, who used it to delineate the three main epochs of human advancement, from savagery through barbarism to civilization.

Although these scholars applied the materialist method to the extent of their ability, their materialism was in many instances crude, inconsistent, and incomplete. This was true even of Morgan, who, as Engels wrote, had rediscovered in his own way the materialist interpretation of history which Engels and Marx had elaborated forty years before. For example, while Morgan classified the main epochs of social development according to the progress made in producing the means of subsistence, in certain places he ascribes the development of institutions and culture to the unfolding of mental seeds: "social and civil institutions, in view of their connection with perpetual human wants, have been developed from a few primary germs of thought" (*Ancient Society,* p. vi).

Despite their deficiencies, the aims and methods of the classical nineteenth-century school were fundamentally correct and bore rich fruit. Their weaknesses have been picked out and exaggerated by their opponents today, not in order to correct them and then probe more deeply into the evolution of humanity, but to exploit them as a means of discrediting the positive achievements and the essentially correct method of the classical anthropologists.

The Reaction

Around the turn of the century new tendencies began to assert themselves in the field of anthropology. These were marked by a growing aversion to the main ideas and methods of the classical school and by a consequent regression in the theoretical level of the science itself. Since then the representatives of these reactionary tendencies have acquired an almost undisputed ascendancy in academic circles, crowding out the doctrines of their predecessors.

Two of the principal currents of thought in this sweeping reaction are the "diffusionist" and the "descriptionist," or "functionalist," schools. Disciples and students of these two tendencies, or of combinations of them, furnish the bulk of the contributors to the Wenner-Gren compilation.

The diffusionists focus their attention upon the beginnings of civilization. Sir G. Elliot Smith, anatomist and leading figure of this school, asserts that "Egypt was the cradle, not only of agriculture, metallurgy, architecture, shipbuilding, weaving and clothing, alcoholic drinks and religious ritual, the kingship and statecraft, but of civilization in its widest sense" (*In the Beginning*, p. 26). The fundamental institutions of civilization spread from that innovating center, with minor accretions and modifications, throughout the world.

Whether or not Egypt was the sole source of all the inventions, as claimed by Smith, the transmission or diffusion of achievements from one people to another is an undeniable factor in the historical process. However, the study of diffusion is no substitute for the analysis of the entire range of social evolution, which covers a far broader field in time and space than this school is willing to survey. Anthropology is, in fact, primarily concerned not with civilized but with savage, or precivilized, society—before agriculture, metallurgy, etc., were born. The diffusionists skip over the most decisive epoch of social evolution, that period from the origin of human society to the threshold of civilization. They shrink from examining the evolution of precivilized life or arranging these stages in any definite historical order.

The pure descriptionists, who dignify their position with the name of "functionalism," proceed without any unified theory of the historical process whatsoever. Their writings have little more theoretical foundation or historical framework than a Boy Scout manual's directions for making Indian objects or imitating Indian dances. Many of them deny that it is necessary, useful, or possible to arrive at any overall view of the course of social development.

This descriptionist current has been best represented by the Franz Boas school in the United States and the Radcliffe-Brown school in England. Having rejected any general view of social evolution, they limit themselves to the study of the cultures and customs of separate peoples and groups. They describe a group's characteristics and occasionally compare or contrast them with one another or with civilized society.

A number of these twentieth-century field investigators have, it is true, brought forth additional important findings that have contributed to the stockpile of materials regarding primitive life. But they view this material in a disconnected way and leave it in an uncoordinated condition. They restrict their views to the framework of each given fragment, and the furthest they go in theoretical interpretation is to try to classify these diverse segments of society into different categories.

Their sole aim is to demonstrate that a variety or diversity of cultures exists and has always existed. They do not even approach the problem—much less answer it—of the specific place these diverse developments occupy and have occupied in the march of human history. They deny that any institution or feature of society is inherently more primitive or advanced than any other. They provide no unifying thread, no guiding line, no definitive acquisitions and advances from one stage to the next in a progressive process of evolution. Nor do they investigate what forces brought about the particular characteristics of each successive level of social development.

By casting aside the theoretical heritage of the classical school, these anthropologists have reduced their science to a patchwork of unrelated facts and data. In place of the historical method, involving a dynamic view of the whole compass of social development, they have substituted a static and purely descriptive approach. This has not only retarded the growth of the science but thrown it back to an infantile theoretical level.

Scientific knowledge progresses from the elementary stage of description and classification of separate phenomena to the more advanced stage of uncovering their organic affiliations and historical interconnections. To the measure of their ability, the pioneer school of anthropologists, employing the evolutionary method, had already proceeded to this higher theoretical stage. But the academic schools which arose in reaction against them reversed this progressive course and slid back to a more primitive level.

Materialism Abandoned

This retrogression arose directly out of the abandonment of the materialist outlook and aims of the classical school. As a rule, the twentieth-century academicians are unwilling and unable to relate the social and cultural institutions of primitive peoples to the economic base upon which they are founded. They deny that the productive forces and activities are decisive in shaping these cultural features. They proceed as though the cultural superstructure developed apart from, and even in opposition to, the technological and productive foundations.

By divorcing culture from its economic roots, some of these anthropologists come to the most absurd conclusions. Elliot Smith, for example, locates the key to human progress not in the advancements made in producing the means of life but in a particular mode of preserving corpses: "It is no exaggeration to claim that the ideas associated with the practice of the embalmer's art have been the most potent influence in building up both the material and spiritual elements of civilization" (*In the Beginning*, p. 51).

The end product of this retrogressive movement is the fashionable psychological and psychiatric approach—latest offspring of the functional school. Margaret Mead, E. Sapir, Ruth Benedict, and other students of Boas are the principal representatives of this new current. In place of the objective material forces and factors which determine the structure and evolution of society, they put forward superficial and arbitrary observations on the different psychological reactions and behaviors of primitive groups. In place of the historical interactions between the developing productive forces and the cultural institutions which spring from them, they substitute the peculiarities of the individual personality.

Margaret Mead, in the Wenner-Gren compilation, locates the key to the differences among cultures not in their different productive and social forces but in the different kinds of weaning and toilet training given to children. Why and how these secondary cultural features arose and evolved she does not explain. The whole fuctionalist school, including its psychological branch, regards "culture" as something disembodied and dematerialized, plucked at will by men out of thin air through inexplicable impulse or caprice.

Leslie A. White, chairman of the Department of Anthropology at the University of Michigan, is one of the few contemporary scholars who has stubbornly refused to abandon the materialist procedures of the Morgan-Tylor school. The most vigorous American critic of the Boas / Radcliff-Brown tendencies, he describes their antimaterialism as follows:

> A few decades ago culture was very real, tangible and observable to anthropologists. They went out to preliterate peoples, saw and collected tools, clothing, ceremonial paraphernalia, utensils and ornaments; they observed people *doing* things—grinding seeds, practicing circumcision, burying prayersticks, chewing betel; they observed expressions of conventional sentiments—a loathing for milk, respect for the mother's brother, a fear of ghosts; they discovered the knowledge and belief of the people. All of this was once as real and tangible to the enthnologist as to the native himself. In recent years, however . . . culture has become an abstraction, intangible, imperceptible, and all but unreal to many anthropologists. . . . What was once a distinct class of real, observable, tangible phenomena, the subject matter of a special science, has now been conjured almost out of existence! [*Philosophy for the Future,* pp. 359-60.]

Flight from Evolutionism

The antimaterialism of the reactionary school is accompanied by its antievolutionism. It is so obvious that stone tools preceded metal tools and food-gathering preceded agriculture and stock-breeding that it is difficult to disclaim evolution altogether. The antievolutionists are obliged to admit that there has been some evolution in technology. But this is as far as they will go in admitting the historical reality.

Above all they deny that social institutions and culture are progressively transformed along with the economic bases of society. They expressly or implicitly deny that the successive social epochs can be delineated through the growth and development of the material forces of production. As a result they not only divorce the cultural superstructure from its material base but flee altogether from any unified and comprehensive conception of historical evolution.

Their chief target for attack is Morgan's thesis of the three main ethnic periods of social evolution: from savagery through barbarism to civilization. Morgan had derived from the changing productive forces at each successive level the changes in the social institutions which flowed from them. He had demonstrated that such fundamental features of civilization as private property and the state did not exist in savagery and only emerged in undeveloped form in barbarism. By the same token, the modern cultural institutions of marriage, the individual family, and the subjugation of women are also less developed the farther back we probe into history. In the epoch of savagery they were nonexistent.

The reactionary anthropologists ridicule and reject these findings of Morgan along with his materialist and evolutionist method. In the Wenner-Gren compilation, Morgan's sequence of ethnic stages in social advancement is relegated to the scrap heap as "out of date," and "mid-Victorian." According to J. Grahame D. Clark, the English archaeologist, Morgan's scheme of determinate ethnic stages is no longer even "respectable." He writes:

> Now it would be ridiculous at this time of day to apportion praise or blame to Morgan, Tylor and the rest; the mid-Victorian anthropologists were confronted by an immense void . . . they merely did what any other scientists would have done under similar circumstances—they plugged the gap with hypotheses . . . their stages were hypothetical. . . . One may legitimately insist, though, that hypothetical prehistory, useful as it may have been 70 or 80 years ago, has long ceased to be respectable [*Anthropology Today,* p. 345].

It is significant that although the "immense void" has been

filled to the brim with further data and documentation during the past seventy or eighty years, Morgan's opponents have not presented any adequate substitute for his theory of social evolution. Having annihilated the positive framework of social evolution developed by the nineteenth-century school, and unable to provide any alternative of their own, the modern schools are manifestly bankrupt in theory and method. Leslie White aptly depicted them as follows:

> In addition to being anti-materialist, they are anti-intellectualistic or anti-philosophic—regarding theorizing with contempt—and anti-evolutionist. It has been their mission to demonstrate that there are no laws or significance in ethnology, that there is no rhyme or reason in cultural phenomena, that civilization is—in the words of R. H. Lowie, the foremost exponent of this philosophy—merely a "planless hodgepodge," a "chaotic jumble" [*Philosophy for the Future*, pp. 367-68].

In truth, the hodgepodge and jumble exist not in the social and cultural phenomena but in the minds and methods of Lowie and his school. Whereas the pioneer anthropologists had sought, and had succeeded to a large degree, in making order out of chaos, the modern academicians have *introduced* chaos into the previously established order. The more materials they accumulate the more narrow their views become. The study of anthropology has become disjointed and jumbled in their hands—and in their students' heads.

Piecemeal Evolutionists

Some contributors to the Wenner-Gren symposium display considerable uneasiness about the absence of any general line of development in primitive history and try to find one. Julian H. Steward, who was assigned the theme of "Evolution and Process," speaks for this group which seeks some middle ground between the classical evolutionists and the modern unabashed antievolutionists. In a subsequent publication which fully develops the ideas in his contribution to the Wenner-Gren book, Steward exposes the unscientific procedures of the "particularists":

> Reaction to evolutionism and scientific functionalism has very

nearly amounted to a denial that regularities exist. . . . It is considered somewhat rash to mention causality, let alone "law," in specific cases. Attention is centered on cultural difference, particulars, and peculiarities, and culture is often treated as if it developed quixotically, without determinable causes, or else appeared full-blown [*Theory of Culture Change*, p. 179].

At the same time Steward ranges himself with the particularists against the advocates of universal evolution, on the specious ground that their generalizations fail to explain particular phenomena:

Universal evolution has yet to provide any very new formulations that will explain any and all cultures. The most fruitful course of investigation would seem to be the search for laws which formulate particular phenomena with reference to particular circumstances [*Anthropology Today*, p. 325].

What Steward is saying in effect is: "To be sure, the world is not flat. However, neither is it quite as round as most people think. Therefore, let us regard it as a flat world with some rounded portions."

According to his own statement, Steward restricts his historical search to "parallels of limited occurrence instead of universals." For example, he and some other American anthropologists sketch out a series of stages in the development of societies on the threshold of civilization, such as Egypt, Mesopotamia, China, Middle America, and the Central Andes. But these parallel lines are never brought together as aspects of a continuous process of social evolution from the lowest stage of savagery up to the threshold of civilization. The particular segments remain disconnected fragments without essential relationship to a general historical framework. Leslie White describes this as piecemeal evolution:

Dr. Steward wants his evolution piecemeal. He wants evolution in restricted areas and in restricted segments. If, however, evolutionist processes and evolutionist generalizations can be made in a number of independent situations and regions, why cannot generalizations be made for evolution as a whole? . . . I notice a rather curious conflict or contradiction of motives in Dr. Steward's scientific work. On the one hand, he seems to be very much interested in generalizations and strives to reach them. On the other hand, he anchors himself to the particular, to the local, or to the restricted, which, of course, tends to inhibit the formulation of broad generalizations [*Philosophy for the Future*, p. 71].

Steward accepts the epoch of civilization (it involves "a less sweeping generalization"), but rejects the two earlier epochs of social development because "they fail to recognize the many varieties of local trends." He then pinpoints the issue upon which he bases his rejection: the proposition that the matriarchy preceded the patriarchy and represents a definite stage in social evolution:

> The inadequacy of unilinear evolution lies largely in the postulated priority of matriarchal patterns over other kinship patterns and in the indiscriminate effort to force the data of all precivilized groups of mankind, which included most of the primitive world, into the categories of "savagery" and "barbarism" [*Anthropology Today*, p. 316].

But the issue goes even deeper than the historical priority of the matriarchy. Morgan and others of the classical school observed that wherever matriarchal vestiges were found, there was also clear evidence that matriarchal society was collectivist and egalitarian. It is this combination of the high position of women and an egalitarian society that lies at the bottom of the stampede away from evolutionism. Basically, it is a flight from Marxism.

Fear of Marxism

In the field of anthropology, as in other fields, a consistently evolutionist and materialist method of thought has revolutionary implications. Unwittingly, the classical anthropologists had verified and supported Marxism as the most scientific system of thought. The science of anthropology did not originate with the historical materialists, but the creators of Marxism drew upon the materials provided by the nineteenth-century anthropologists to extend their own historical reach and substantiate their materialist interpretation of history. They drew out to their logical conclusions the sharp contrasts between capitalism, the highest form of class society, and primitive, or preclass society. These conclusions are set forth in the renowned work by Engels, *Origin of the Family, Private Property, and the State*, which appeared in 1884.

The reactionary flight from materialism and evolutionism arose out of the effort to counter this challenge of Marxism. But in the process of disowning the views of the Marxists, they were obliged to also turn against the pioneers in their own field of science.

The repudiation by these modern anthropologists of the principles and methods of their pioneer predecessors had its precedent and parallel in the rejection by academic economists of their classical bourgeois predecessors, from Smith to Ricardo. The labor theory of value, which was taken over from the classical economists and developed by the Marxists, produced the revolutionary conclusions of *Capital.* Subsequent bourgeois economists, recoiling from these conclusions, found it expedient to dump, along with them, the positive achievements of their own predecessors. The same thing has happened in anthropology. The Marxists connected Morgan's findings with the conclusion that just as primitive collectivism had been destroyed by class society so, in turn, would class society be replaced by the new higher stage of socialism. The modern reactionary school, in flight from this conclusion, was obliged not only to oppose the Marxists but to reject their own predecessors, whose findings substantiated this view.

There is no ambiguity on this score in the Wenner-Gren compilation. Grahame Clark explains why the Morgan-Tylor school must be cast out, along with Marxism:

> . . . Marxists find in archaeology a means of recovering what they hold to be tangible evidence for the validity of the dogma of the materialist interpretation of history. . . . What is quite sure is that Marxist dogma is no more valid as a substitute for archaeological research than were the speculations of Victorian ethnologists. Both are equally out of date [*Anthrolopology Today*, p. 346].

Julian Steward likewise explains why a consistently evolutionary position is intolerable:

> The Marxist and Communist adoption of the 19th century evolutionism, especially of L. H. Morgan's scheme, as official dogma (Tolstoy 1952) has certainly not favored the acceptability of scientists of the Western nations of anything labeled "evolution" [ibid., p. 315].

Here, then, is the underlying reason for the antimaterialism and antievolutionism of so many contemporary anthropologists. The reactionary school has become predominant because it has accommodated itself to ruling-class prejudices and dogmas and assumed the obligation of stamping out the spread of revolutionary conclusions.

The Road Ahead

Darwin provided a solid foundation for biology and directed it along correct lines by explaining how animal species originated and evolved from one order to another. Until the science of anthropology likewise discovers the secrets of the social cradle of humanity, it lacks such a solid foundation. The road ahead for anthropology today lies precisely in this deeper penetration into our most remote past, above all, at that critical juncture where the first social horde emerged from the animal world.

This central problem was not neglected by the nineteenth-century pioneer school. On the contrary, their serious and systematic research provided a sound point of departure. Morgan had detected that the gens, or clan system, arose as the universal and fundamental form of primitive organization. He also detected that the gens had been preceded by a cruder, more unfinished form of social organization based upon "male and female classes." This "classificatory system" was subsequently subsumed into the gens system. The Scotsman J. L. McLennan called attention to the importance of the strange code of social and sexual rules which has been voluminously discussed under the various headings of totemism, taboo, and exogamy. W. H. R. Rivers understood that decisive clues were contained in that peculiarity of the gens system called "dual organization." Sir James Frazer and others assembled monumental researches on these and other bewildering phenomena, but their meaning in the formation and rise of the primitive gens system remained enigmatic. The principal merit of these pioneers was not the answers they provided but rather the materials they assembled, the penetrating observations they made, and the questions they posed. Their work was and still remains the precondition for the solution of social origins.

Despite this wealth of material and the crucial importance of the subject, the question of social origins is neglected in the Wenner-Gren inventory. Not only have they ceased to follow the trail begun by the nineteenth-century investigators but they have ignored the key theoretical contributions to this study already available.

In the nineteenth century, Engels sought the decisive social factor which had lifted humanity out of the animal world. The Marxists had already established that all society, from lower to higher stages, moved forward with the advances made in labor techniques and production. But Engels called attention to the fact that labor was the key to human beginnings, and the birth of labor activities was simultaneously the birth of humanity. This labor theory of social origins is set forth in Engels's essay "The Part Played by Labour in the Transition from Ape to Man."

Some fifty years later, in 1927, Robert Briffault's work entitled *The Mothers* was published, providing the biological link to this proposition. He demonstrated that maternal functions were the indispensable biological basis for the first laboring activities and social cooperation. Earlier investigators had established that the matriarchy represented a definite stage in social evolution. Briffault went a step further than this and showed *why* the matriarchal form was the necessary and unavoidable first form of society. He called this the matriarchal theory of social origins.

The theories of Engels and Briffault dovetail. If, as Engels has explained, labor was the central factor in transforming our branch of the anthropoid species into humanity, and if, as Briffault has shown, the females were the pioneers and leaders in labor, it follows that women-as-laborers provided the main living force in developing the first social horde.

At all stages, Marxists have pointed out, society is founded upon the twin pillars of production and reproduction. In civilized society these two functions have been divided between the sexes. The production of new life remains the sphere of the women, while the production of the means of life is primarily in the hands of the men. But at the beginning of human time, and for more than 90 percent of subsequent history, women were not only the *procreators* but also the *principal producers* of the means of subsistence. What Briffault explained was that because women produced and cared for new life, they became the first producers of the means of life.

Thus, the historical primacy of the matriarchy, which is rejected by the academic schools, actually holds the key to solving the basic question of social origins. There are still many unanswered questions, among them the question of why the first society was not only matriarchal but collectivist in productive and social relations. But the solution to all the problems connected with social beginnings must start with the indispensable guidelines provided by Engels and Briffault. Equally important is the need to restore the materialist and evolutionist methods of the nineteenth-century classical school. Enriched by the more extensive data available today and aided by Marxist historical materialism, anthropology can not only be brought out of its stagnation and sterility but elevated to a new and higher level.

Glossary

AFFILIATED CLANS: A general category including both linked (parallel) clans and allied (cross-cousin) clans.

ANTHROPOIDS: Human-like apes out of which the hominids evolved.

ANTHROPOLOGY: The science of the prehistoric evolution of human society.

ANTHROPOMORPHISM: Ascribing human traits to things not human.

ARTIFACT: Object processed by human labor.

BARBARISM: The second and higher level of social evolution, after savagery, with an economy based on agriculture and stock-raising.

BIOLOGY: The science of life and living organisms.

BLOOD REVENGE: A primitive system of reciprocal punishment for deaths. Also called vengeance fighting and blood feud.

CIVILIZATION: The third and present stage of social evolution, marked by the emergence of private property, class divisions, and the patriarchal family.

CLAN: The unit of primitive society, as contrasted with the family, the unit of civilized society. The primitive clan is always a maternal clan composed of social sisters and brothers; an exogamous unit within a tribe.

CLASSIFICATORY KINSHIP: The clan system of social kinship, binding together a large group of people without reference to genetic (family) ties.

COMMODITY EXCHANGE: The transfer of commodities from one owner to another for commodities of a different kind or money; a commercial transaction that is the opposite of interchange, or reciprocal gifts.

CROSS-COUSINS: Men and women of the same generation belonging to opposite (and formerly hostile) groups; the men of the two sides having fraternal relations with each other as "brothers-in-law," and the women and men of the two sides being actual or potential mates.

DIVIDED FAMILY: The matrifamily, in which the mother's brother predominates over the mother's husband in relation to the children.

DUAL ORGANIZATION: Reciprocal relationship between two exogamous communities (moieties) for the interchange of food, mates, and friendly social relations.

ENDOGAMY: A reciprocal relationship between two exogamous communities (moieties) for the interchange of food, mates, and friendly social relations. A feature of the dual organization.

ESTRUS: Period during which the female of many mammalian species is sexually receptive and capable of conceiving.

EVOLUTION, SOCIAL: The theory that society has passed through successive stages of development from lower to higher.

EXOGAMY: The "marrying out" rule, which is also a rule of "hunting out." Men belonging to a clan kinship group must obtain their food and mates outside their own territory.

FATHER-FAMILY: The patriarchal family of civilized society, consisting of a father standing at the head of his wife or wives and their children.

FOSSIL: An organic object that has been prevented from perishing by solidification into stone.

FRATRIARCHY: The brotherhood of men bound together by their social kinship ties; the male counterpart of the matriarchy, a sisterhood of women.

FRATRILINEAL KINSHIP: The male corollary of matrilineal kinship.

GIFT-GIVING: The interchange of food and other things between groups to create fraternal and matrimonial relations.

HOMINIDS: Various beings above the apes but below *Homo sapiens;* subhumans.

INCEST: Sexual intercourse between the genetic relatives in a family.

INTERCHANGE: Reciprocal social relations involving gift-giving; distinct from barter and commodity exchange, which evolved later.

KINSHIP: Relatives by maternal clan ties in primitive society; by family/genetic ties in civilized society.

KULA: Trobriand term for their interchange system.

MATRIARCHY: The maternal communal clan system of social organization that preceded patriarchal civilized society.

MATRIFAMILY: The first form of the family prior to the patriarchal family, also called the "pairing family." See also Divided Family.

MATRILOCAL MARRIAGE: Residence of a cohabiting pair in the community of the wife.

MOIETY: One of two sides of a tribe. Each of the two sides is a phratry.

NON-KIN: Strangers with whom primitive peoples have no kinship or social relations.

PAIRING FAMILY: Morgan's term for the matrifamily.

PARALLEL COUSINS: Men and women of the same generation belonging to different but linked clans. The men are "brothers" and the women are "sisters," and sexual relations between them are forbidden.

PATRIARCHY: The supremacy of the father and the male sex in general in social and family life.

PATRILINEAL KINSHIP: Acknowledgment of paternal ties between a child and its mother's husband.

PATRILOCAL MARRIAGE: Residence of a cohabiting pair in the community of the husband and his maternal kin.

PHRATRY: A number of linked (parallel) clans comprising one side or moiety of a tribe.

POTLATCH: Northwest Coast Indian ceremony featuring food and gift interchange.

PRIMAL HORDE: The earliest social group emerging from primate life.

PRIMEVAL PERIOD: The lower stage of savagery.

PRIMATOLOGY: The study of primates, both apes and monkeys.

PRIMITIVE: Term usually applied to the upper stage of savagery, although the first stage of barbarism is also considered primitive.

PRIVATE PROPERTY: Wealth possessed by individuals or individual families, as opposed to the communal property of primitive society.

SAVAGERY: The earliest, most primitive stage of social evolution, with an economy based on hunting and gathering.

SOCIOBIOLOGY: A new term coupling sociology with biology. As interpreted by E. O. Wilson and others, it is based on a crude form of biological determinism.

SOCIOLOGY: The science of the origin and evolution of social organization.

TABOO: Prohibition or quality of being prohibited.

TOTEMISM: The earliest system of social regulation, based on totem kinship and taboos.

TRIBE: An endogamous community composed of two exogamous phratries or sides, each composed of several linked clans.

VEVE: Melanesian term for maternal clan.

Bibliography

Altmann, Stuart A. *Social Communication Among Primates.* Chicago: University of Chicago Press, 1967.

Ardrey, Robert. *African Genesis: A Personal Investigation Into the Animal Origins and Nature of Man.* New York: Dell Publishers, 1963 (1961).

—*The Territorial Imperative.* New York: Dell Publishers, 1966.

Bachofen, J. J. *Myth, Religion and Mother Right.* Translated by Ralph Mannheim. Princeton: Princeton University Press, 1967 (1861).

Boulding, Kenneth E. "Am I A Man Or A Mouse—Or Both?" In *Man and Aggression,* M. F. Ashley Montagu, ed. New York: Oxford University Press, 1969 (1968).

Briffault, Robert. *The Mothers: A Study of the Origin of Sentiments and Institutions.* 3 vols. New York: Macmillan; London: Allen and Unwin; 1952 (1927).

Carrighar, Sally. "War Is Not in Our Genes." In *Man and Aggression,* M. F. Ashley Montagu, ed. New York: Oxford University Press, 1969 (1968).

Childe, V. Gordon. *Man Makes Himself.* New York: New American Library, 1951 (1936).

—*What Happened in History.* Harmondsworth, Middlesex, England: Penguin Books, 1960 (1942).

Clark, J. Grahame D., "Archeological Theories and Interpretation: Old World." In *Anthropology Today: An Encyclopedic Inventory,* A. L. Kroeber, ed. Chicago: University of Chicago Press, 1953.

Clark, W. E. LeGros. *History of the Primates: An Introduction to the Study of Fossil Man.* 2d ed. London: British Museum, 1950.

Cockburn, Alexander. Review of *Sociobiology: The New Synthesis.* In the *Village Voice,* July 28, 1975.

Darwin, Charles. *The Descent of Man.* 2d ed., revised. New York: Burt, 1874 (1871).

Eaton, C. Gray. "The Social Order of Japanese Macaques." *Scientific American,* October 1976.

Eggan, Fred. "Aboriginal Sins." Review of *Remarks and Inventions: Sceptical Essays about Kinship,* by Rodney Needham. In *London Times Literary Supplement,* December 13, 1974.

Emlen, John T., Jr., and Schaller, George. "In the Home of the Mountain Gorilla." In *Primate Social Behavior,* Charles H. Southwick, ed. Princeton: D. Van Nostrand Co., 1963.

Engels, Frederick. *The Origin of the Family, Private Property, and the*

Note: Where a later edition is listed, the date of the original edition is given in parentheses.

Howard Haymes's bibliography follows Evelyn Reed's.

State. Introduction by Evelyn Reed. New York: Pathfinder Press, 1972.
—"The Part Played by Labour in the Transition from Ape to Man." In *The Origin of the Family, Private Property, and the State*. Introduction by Evelyn Reed. New York: Pathfinder Press, 1972.

Fox, Robin. *Encounter with Anthropology*. New York: Dell Publishers, Laurel edition, 1975 (1968).
—*Kinship and Marriage*. Harmondsworth, Middlesex, England: Penguin Books, 1967.
Galdikas-Brindamour, Biruté. "Orangutans, Indonesia's 'People of the Forest.'" *National Geographic,* October 1975.
Goodall, Jane Van Lawick. *In the Shadow of Man*. Boston: Houghton Mifflin Co., 1971.
Gorer, Geoffrey. "Ardrey on Human Nature: Animals, Nations, Imperatives." In *Man and Aggression,* M. F. Ashley Montagu, ed. New York: Oxford University Press, 1969 (1968).
Gould, Stephen Jay. "Posture Maketh the Man." *Natural History,* November 1975.
Hahn, Emily. *On the Side of the Apes*. New York: Thomas Y. Crowell, 1968.
Hall, K. R. L. "Behavior and Ecology of the Wild Patas Monkeys in Uganda." In *Primates,* Phyllis C. Jay, ed. New York: Holt, Rinehart & Winston, 1968.
—"Some Problems in the Analysis and Comparison of Monkey and Ape Behavior." In *Classification and Human Evolution,* S. L. Washburn, ed. Chicago: Aldine Publishing Co., 1963.
—"Tool-Using Performances as Indicators of Behavioral Adaptability." In *Primates,* Phyllis C. Jay, ed. New York: Holt, Rinehart & Winston, 1968.
Harris, Marvin. *The Rise of Anthropological Theory*. New York: Thomas Y. Crowell, 1968.
—"Male Supremacy is on the Way Out. It was Just a Phase in the Evolution of Culture." Interview with Carol Travis. In *Psychology Today,* January 1975.
Hoebel, E. Adamson. *Man in the Primitive World*. New York: McGraw-Hill Co., 1949.
Holloway, Ralph. "Territory and Aggression in Man: A Look at Ardrey's 'Territorial Imperative.'" In *Man and Aggression,* M. F. Ashley Montagu, ed. New York: Oxford University Press, 1969 (1968).

Howells, William W. *Mankind So Far*. American Museum of Natural History Science Series. Garden City, N.Y.: Doubleday, 1944.
Jay, Phyllis C., ed. *Primates: Studies in Adaptation and Variability*. New York: Holt, Rinehart & Winston, 1968.
Jones, F. Wood. *Arboreal Man*. London: Arnold & Co., 1926.
Korn, Francis. *Elementary Structures Reconsidered: Lévi-Strauss on Kinship*. Berkeley and Los Angeles: University of California Press, 1973.
Kroeber, A. L., ed. *Anthropology Today: An Encyclopedic Inventory*. Chicago: University of Chicago Press, 1953.

Lancaster, Jane Beckman. "Stimulus Response." *Psychology Today,* September 1973.
Leach, E. R. *Rethinking Anthropology.* London: Athlone Press, 1966.
Leacock, Eleanor. "The Changing Family and Lévi-Strauss, or Whatever Happened to Fathers?" *Social Research,* summer 1977.
—and Nash, June. "Ideologies of Sex, Archetypes and Stereotypes." In *Issues in Cross-Culture Research.* Annals of the New York Academy of Sciences, vol. 285, 1977.
Lévi-Strauss, Claude. *The Elementary Structures of Kinship.* Revised ed. Boston: Beacon Press, 1969 (1949).
—*The Savage Mind.* Chicago; University of Chicago Press, 1966.
—*Structural Anthropology.* New York: Basic Books, 1967.
—*Totemism.* Boston: Beacon Press, 1963 (1962).
Linton, Ralph. *The Study of Man* (excerpts). In *Sex Differences,* Patrick C. Lee and Robert S. Stewart, eds. New York: Urizen Books, 1976.
—*The Tree of Culture.* New York: Alfred A. Knopf, 1955.
Lorenz, Konrad. *On Aggression.* New York: Harcourt, Brace & World, 1966.
Lowie, Robert H. *The History of Ethnological Theory.* New York: Farrar & Rinehart, 1937.
Montagu, M. F. Ashley, ed. *Man and Aggression.* Oxford: Oxford University Press, 1969.
—"The New Litany of 'Innate Depravity.'" In *Man and Aggression,* M. F. Ashley Montagu, ed. Oxford: Oxford University Press, 1969.
Morgan, Lewis H. *Ancient Society.* Chicago: Kerr [1877].
Morris, Desmond. *The Naked Ape.* New York: Dell Publishers, 1967.
Needham, Rodney. "Prospects and Impediments" for "The State of Anthropology." *London Times Literary Supplement,* July 6, 1973.
—*Remarks and Inventions: Sceptical Essays about Kinship.* London: Tavistock, 1974.
Oakley, Kenneth P. *Man the Tool-Maker.* London: British Museum, 1950.
Penniman, T. K. *A Hundred Years of Anthropology.* London: Gerald Ducksworth & Co., 1952.
Reed, Evelyn. *Woman's Evolution: From Matriarchal Clan to Patriarchal Family.* New York: Pathfinder Press, 1975.
Reining, Priscilla, ed. *Kinship Studies in the Morgan Centennial Year.* Anthropological Society of Washington, 1972.
Rohrlich-Leavitt, Ruby. *Peaceable Primates and Gentle People.* New York: Harper & Row, 1975.
Sahlins, Marshall D. and Service, Elman R., eds. *Evolution and Culture.* Ann Arbor: University of Michigan Press, 1959.
Sanderson, Ivan T. *The Monkey Kingdom: An Introduction to Primates.* Philadelphia and New York: Chilton Books, 1963.
Schaller, George B. *The Year of the Gorilla.* Chicago: University of Chicago Press, 1971.
Schneider, David M. and Gough, Kathleen, eds. *Matrilineal Kinship.* Berkeley and Los Angeles: University of California Press, 1961.
Scott, J. P. "That Old-Time Aggression." In *Man and Aggression,* M. F. Ashley Montagu, ed. Oxford: Oxford University Press, 1969.

Simpson, George Gaylord. *The Meaning of Evolution.* New Haven: Yale University Press, 1969.
Shapiro, Arnold. "Dr. Strum: Presenting Baboons." In *Westways* magazine, vol. 69, no. 3, March 1977.
Steward, Julian H. "Evolution and Process." In *Anthropology Today: An Encyclopedic Inventory,* A. L. Kroeber, ed. Chicago: University of Chicago Press, 1953.
Strum, Shirley C. "Life with the Pumphouse Gang." *National Geographic,* May 1975.
Tiger, Lionel. "Male Dominance? Yes, Alas. A Sexist Plot? No." *New York Times Magazine,* October 25, 1970.
—*Men in Groups.* New York: Vintage, 1970 (1969).
Tilney, Frederick. *The Master of Destiny.* Garden City, New York: Doubleday, Doran & Co., 1929.
Tylor, Edward Burnett. *Anthropology.* Abridged ed. Ann Arbor: University of Michigan Press, 1970 (1881).
—"On a Method of Investigating the Development of Institutions." In *Source Book in Anthropology,* A. L. Kroeber and T. T. Waterman, eds. New York: Harcourt, Brace, 1931.
Washburn, S. L., ed. *Classification and Human Evolution.* Chicago: Aldine Publishing Co., 1962.
—*Social Life of Early Man.* Chicago: Aldine Publishing Co., 1961.
—and DeVore, Irven. "The Social Life of Baboons." In *Primate Social Behavior,* Charles W. Southwick, ed. New York: D. Van Nostrand Co., 1963.
—and DeVore, Irven. "Social Behavior of Baboons and Early Man." In *Social Life of Early Man,* S. L. Washburn, ed. Chicago: Aldine Publishing Co., 1961.
—and Hamburg, D. A. "Aggressive Behavior in Old World Monkeys and Apes." In *Primates,* Phyllis C. Jay, ed. New York: Holt, Rinehart & Winston, 1968.
Wertheim, W. F. *Evolution and Revolution: The Rising Waves of Emancipation.* London: Penguin Books, 1974.
White, Leslie A. "Ethnological Theory." In *Philosophy for the Future,* Marvin Farber, F. J. McGill, and Roy W. Sellars, eds. New York: Macmillan, 1949.
Wilson, Edward O., *Sociobiology: The New Synthesis.* Cambridge, Mass.: Belknap Press of Harvard University Press, 1975.
Yerkes, Robert M. *Chimpanzees: A Laboratory Colony.* New Haven: Yale University Press, 1943.

Howard Haymes Bibliography

Ti-Grace Atkinson. "The Institution of Sexual Intercourse." New York: Mimeographed by *The Feminist,* 1968.

—"Radical Feminism." New York: Mimeographed, 1969.

Bachofen, J. J. *Myth, Religion and Mother Right.* Translated by Ralph Mannheim. Princeton: Princeton University Press, 1967 (1861).

Bird, Caroline. *Born Female.* New York: David McKay Co., 1970.

Diner, Helen. *Mothers and Amazons.* Garden City, N.Y.: Doubleday & Co., 1965.

Dixon, Marlene. "Why Women's Liberation?" *Ramparts,* November 1969.

Engels, Frederick. *The Origin of the Family, Private Property, and the State.* In *The Woman Question.* New York: International Publishers, 1951.

Figes, Eva. *Patriarchal Attitudes.* New York: Stein & Day, 1970.

Firestone, Shulamith. *Dialectic of Sex: The Case for Feminist Revolution.* New York: Bantam Books, 1971.

Gibbs, James L., Jr. "Social Organization." In *Horizons of Anthropology,* Sol Tax, ed. Chicago: Aldine Publishing Co., 1964.

Jones, Beverly and Brown, Judith. *Toward a Female Liberation Movement.* Boston: New England Free Press, 1968.

Kardiner, Abram and Preble, Edward. *They Studied Man.* New York: World Publishers, 1961.

Marx, Karl and Engels, Frederick. *The Communist Manifesto.* In *Introduction to Contemporary Civilization in the West.* 2 vols., 3d ed. New York: Columbia University Press, 1961.

Millett, Kate. *Sexual Politics.* New York: Avon Books, 1969.

Mitchell, Juliet. "Women: The Longest Revolution." *New Left Review,* November/December 1966.

Morgan, Lewis Henry. *Systems of Consanguinity and Affinity of the Human Family.* In *Readings in Anthropology,* Morton H. Fried, ed. (entitled "General Observations Upon Systems of Relationship"). 2 vols., 2d ed. New York: Thomas Y. Crowell Co., 1968.

Morgan, Robin, ed. *Sisterhood Is Powerful.* New York: Random House, 1970.

Reed, Evelyn. *Problems of Women's Liberation: A Marxist Approach.* New York: Pathfinder Press, 1970.

Sherfey, Mary Jane. "Ancient Man Knew His Place." *New York Times,* November 14, 1972.

Index

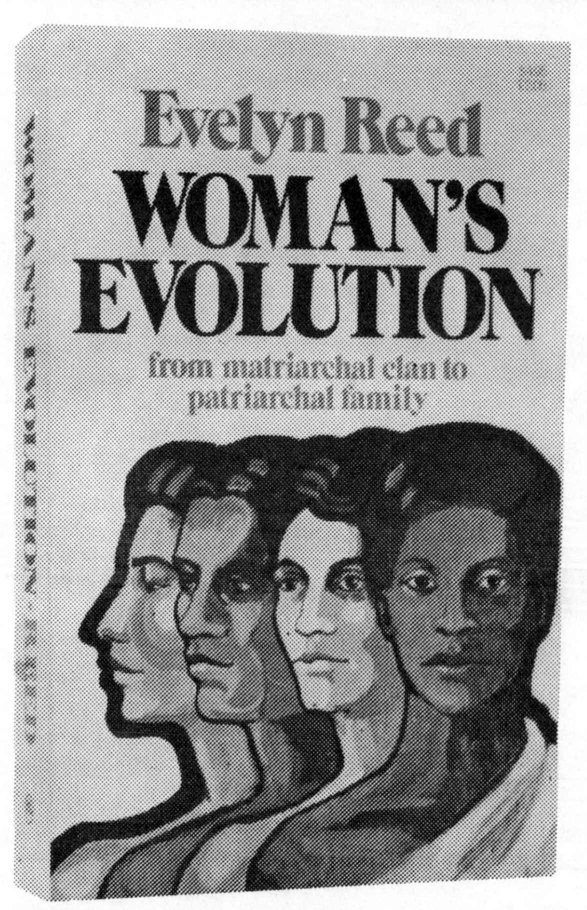
Evelyn Reed
WOMAN'S EVOLUTION
from matriarchal clan to patriarchal family